SpringerBriefs in Stem Cells

More information about this series at http://www.springer.com/series/10206

Tao Cheng

Editor

Hematopoietic Differentiation of Human Pluripotent Stem Cells

 Springer

Editor
Tao Cheng
Institute of Hematology
Tianjin
China

ISSN 2192-8118   ISSN 2192-8126   (electronic)
SpringerBriefs in Stem Cells
ISBN 978-94-017-7311-9   ISBN 978-94-017-7312-6   (eBook)
DOI 10.1007/978-94-017-7312-6

Library of Congress Control Number: 2015945324

Springer Dordrecht Heidelberg New York London

Printed on acid-free paper

Springer Science+Business Media B.V. Dordrecht is part of Springer Science+Business Media (www.springer.com)

Contents

Chapter 1
Generation of Hemangioblasts from Human Pluripotent Stem Cells

Shi-Jiang Lu, Qiang Feng and Robert Lanza

Abstract Hemangioblasts are progenitors with the capacity to differentiate into hematopoietic, endothelial, smooth muscle, and mesenchymal stromal cells and represent an excellent candidate of cell therapy for a variety of human diseases. To realize their clinical potential, first, an efficient and controlled differentiation toward hemangioblasts in a scalable manner, probably in a bioreactor setting, from an unlimited source is required. These cells also need to be generated under animal components and cell-free conditions, or additional regulatory challenges, such as xeno-transplant, will have to be addressed. The ability of human embryonic stem cells (hESCs) and induced pluripotent stem cells (iPSC) to divide indefinitely without losing pluripotency may allow them to serve as an inexhaustible source for the large-scale production of therapeutic cells. In this chapter, we describe a robust system that can efficiently generate large numbers of hemangioblasts from multiple hESCs and iPSC lines under well-defined conditions, which is an important step for future clinical applications, with the potential of developing a GMP-compatible scalable system.

Keywords Human embryonic stem cells · Induced pluripotent stem cells · Hemangioblast · Cell therapy

1.1 Introduction

Hemangioblasts are progenitors with the capacity to differentiate into hematopoietic and vascular (endothelial and smooth muscle cells) and mesenchymal stromal cells [1–5]; thus, they are excellent candidates for cell therapy in both blood and vascular diseases. Myocardial infarction, stroke, coronary artery disease, ischemic limbs caused by diabetes, and diabetic retinopathy are devastating and life-threatening

S.-J. Lu (✉) · Q. Feng · R. Lanza
Ocata Therapeutics, Marlborough, MA, USA
e-mail: sjlu@ocata.com

© The Author(s) 2015
T. Cheng (ed.), *Hematopoietic Differentiation of Human Pluripotent Stem Cells*,
SpringerBriefs in Stem Cells, DOI 10.1007/978-94-017-7312-6_1

ischemic diseases which are primarily caused by vascular dysfunction. The ability to repair vascular damage could have a profound impact on these major diseases afflicting humans. Although adoptive transfer of endothelial precursor cells has been previously shown to restore blood flow and increase capillary density, decreasing limb loss and facilitating recovery from myocardial injury [6, 7], formation and regeneration of functional vasculatures require both endothelial cells (EC) and vascular smooth muscle cells (SMC) [8–10]. Blood vessels are typically composed of two major cell types: the inner endothelium, a thin layer of ECs that separate the blood from tissues, and an outer layer of mural cells (pericytes and vascular SMCs) that protect the fragile channels from rupture and help control blood flow [9]. Although ECs play an essential role in vasculogenesis and angiogenesis, they alone cannot complete the process of vessel growth and development. The formation of a mature and functional vascular network requires communication between ECs and SMCs. Vascular SMCs play a critical role in structural and functional support of the vascular network by stabilizing nascent endothelial vessels during vascular development and blood vessel growth [9]. Hemangioblasts, therefore, will be the ideal cell type for this purpose.

Hemangioblasts have been identified in differentiation cultures of mouse and non-human primate as well as human ESCs and iPSCs [11–19]. We have demonstrated that hemangioblasts derived from hESCs, after injected into animals with spontaneous type II diabetes or ischemia/reperfusion (I/R) injury of the retina, homed to the site of injury and showed robust reparative function of the damaged vasculature. These cells also showed a similar regenerative capacity in NOD/SCID β2–/– mouse models of both myocardial infarction (50 % reduction in mortality rate) and hind limb ischemia, with restoration of blood flow in the later model to near-normal levels [17]. Fluorescent immunocytochemistry showed that the vascular lumens were surrounded by human ECs and SMCs in both diabetic and I/R damaged retinas. Similarly, confocal microscopy confirmed the incorporation of human ECs and SMCs into the lumens of microvessels in the infarcted heart and ischemic limb tissues [4, 17]. These results demonstrate that hemangioblasts generated from hESCs can provide a potentially inexhaustible source of cells for the treatment of human vascular diseases.

However, before hESC and iPSC derivatives can be used in the clinic, it is important to understand the steps involved, as well as the risks and challenges associated with the production of therapeutic products. First, an efficient and controlled differentiation toward the specific lineage cell fate in a scalable manner, probably in a bioreactor setting, is required. Second, derivatives from hESCs or iPSCs need to be generated under animal component/cell-free conditions, or additional regulatory challenges, such as xeno-transplant, will have to be addressed. We have established a simple strategy to efficiently and reproducibly generate hemangioblasts from multiple hESC lines maintained on mouse embryonic fibroblast (MEF) layer [17, 18]; however, massive cell death was observed for hESCs maintained under feeder-free condition after platting for EB formation [18, 19]. We recently adapted a 3D microcarrier system to passage and expand hESCs and iPSCs in liquid culture [19, 20]. After simple replacement of mTeSR medium with EB formation medium, these cells readily formed EBs with minimum cell death, and cells from these EBs readily developed into hemangioblasts after replating under appropriate conditions, which is indistinguishable from hemangioblasts derived from hESCs cultured under 2D conditions [19]. In this chapter, we describe both systems that can efficiently generate large numbers of hemangioblasts from hESCs and iPSCs using well-defined conditions.

1.2   Hemangioblast Generation from hESCs Cultured on MEF [17, 18]

1.2.1   Materials

1. bFGF stock solution: add 1.25 ml of protein-containing medium (use hESC-BM plus 20 % serum replacement) to a vial containing 10 μg of bFGF. This makes 8 mg/ml stock solution. Make 240 μl aliquots and freeze at −20 °C.

2. PMEF growth medium: To a 500-ml bottle of high glucose DMEM, add 6 ml penicillin/streptomycin (100× solution), 6 ml Glutamax-1 (100× solution), and 50 ml fetal bovine serum (FBS), sterilize by 0.22 μm filtration, and store at 4 °C.

3. hESC basal medium (hESC-BM): To a 500-ml bottle of KO-DMEM, add 6 ml penicillin/streptomycin, 6 ml Glutamax-1, 6 ml nonessential amino acids (NEAA), (100× solution), and 0.6 ml β-mercaptoethanol (1000× solution) and store at 4 °C.

4. hESC growth medium (hESC-GM): To 200 ml of hESC-BM, add 40 ml knockout serum replacement, 240 μl of human LIF for 10 ng/ml, and 240 μl of bFGF for 8 ng/ml. Sterilize by 0.22 μm filtration and store at 4 °C. Note: Primate ES growth medium (Cat # RCHEMD001) from ReproCell worked well for culturing human ESCs on MEF.

5. Gelatin (0.1 %): dissolve 0.5 g of gelatin (from porcine skin, Sigma) in 500 ml of warm (50–60 °C) Milli-Q water. Cool down to room temperature and sterilize by 0.22 μm filtration.

6. Mitomycin C: add 2 ml of sterile Milli-Q water to a vial (2 mg) of lyophilized mitomycin C to make 1 mg/ml stock solution. The solution is light-sensitive and is good for 1 week at 4 °C (Note: Some batches of mitomycin C appear to become very light in color and form an insoluble precipitate, so always check the intensity of the color (should be deep purple) and for the presence of the precipitate. Do not use if different from freshly prepared stock. PMEFs treated with such discolored mitomycin C seem to proliferate and show intensive labeling with BrdU. An alternative to mitomycin C treatment is to irradiate PMEFs (suspended in 5 ml medium) in a blue polypropylene 15-ml conical tube with a dose of 3500–4000 rad).

7. VEGF solution: add 1 ml of PBS (Ca^{++}, Mg^{++}-free) with 1 % BSA to a vial containing 50 µg of human recombinant VEGF$_{165}$ (R&D Systems). Make 100 µl aliquots and freeze at −20 °C. This makes a 50 µg/ml stock solution.

8. BMP-4 solution: add 1 ml of PBS (Ca^{++}, Mg^{++}-free) with 1 % BSA to a vial containing 10 µg of human recombinant BMP-4 (R&D Systems). Make 100 µl aliquots and freeze at −20 °C. This makes 10 µg/ml stock solution.

9. Embryoid body (EB) formation medium-1: transfer 20 ml of Stemline II hematopoietic stem cell expansion medium (Sigma) into a 50-ml tube and add 0.2 ml of penicillin/streptomycin, 20 µl of VEGF, and 100 µl of BMP-4. Sterilize by 0.22 µm filtration and store at 4 °C up to a month.

10. EB formation medium-2: transfer 20 ml of Stemline II hematopoietic stem cell expansion medium into a 50-ml tube and add 0.2 ml of penicillin/streptomycin, 20 µl of VEGF, 100 µl of BMP-4, and 100 µl of bFGF (8 µg/ml). Sterilize by 0.22 µm filtration and store at 4 °C up to a month.

11. Flt3 ligand solution: add 1 ml of PBS (Ca^{++}, Mg^{++}-free) with 1 % BSA to a vial containing 25 µg of human recombinant Flt3 ligand (R&D Systems). Make 100 µl aliquots and freeze at −20 °C. This makes a 25 µg/ml stock solution.

12. TPO solution: add 1 ml of PBS (Ca^{++}, Mg^{++}-free) with 1 % BSA to a vial containing 25 µg of human recombinant TPO (R&D Systems). Make 100 µl aliquots and freeze at −20 °C. This makes a 25 µg/ml stock solution.

13. Blast cell growth medium (BGM): To a 100-ml bottle of serum-free hematopoietic CFC medium (Stem Cell Technologies, Cat # 4436 or 4536), add 1 ml of penicillin/streptomycin, 1 ml of Ex-Cyte growth enhancement media supplement (Millipore), 100 µl of VEGF, 200 µl of Flt3 ligand, 200 µl of TPO, and 250 µl of bFGF. Mix well by shaking. Stand for 5–10 min and then aliquot 2.5–3 ml/tube by using a 3- to 5-ml syringe with a 16- or 18-gauge needle. Store at −20 °C.

1.2.2   Methods

1.2.2.1   Prepare Primary Mouse Embryo Fibroblasts (PMEFs) from 12.5 dpc CD-1 Mouse Embryos [21]

Note: Mitomycin C-treated MEFs from several companies worked well in our hand.

1. Plate and grow early passage PMEFs (<P5) in 150-mm tissue culture plate to confluency.

2. Add 10 mg/ml mitomycin C to the media and incubate at 37 °C for 3 h.

3. Rinse mitomycin C-treated PMEFs 3 times with PBS. Add 4 ml of 0.05 % trypsin/0.53 mM EDTA and incubate at 37 °C for 2–4 min and then add 10 ml of PMEF medium to inactivate trypsin.

4. Collect inactivated PMEFs by centrifugation at 1000 rpm (210 g) for 5 min.

5. Count and plate inactivated PMEFs onto 0.1 % gelatin-coated 6-well plates at a density of 7.5×10^5 cells/well (in a 6-well plate) in PMEF medium. PMEF feeders should be prepared at least 1 day before culturing hES cells and remain suitable up to 5 days.

1.2.2.2 Culture of Undifferentiated hESCs [22]

1. Add 2 ml of hESC-GM to PMEF plates and equilibrate in the CO_2 incubator for 30 min or longer before plating hESCs for helpful recovery of hESCs.
2. Take a vial of frozen hESCs out from liquid nitrogen and immediately put in a 37 °C water bath, constantly agitating the vial while ensuring that the neck of the vial is above the water level.
3. When last sliver of ice in vial remains (after about a minute in 37 °C), spray the vials with 70 % isopropanol, using a 1-ml pipetman, and add warm hESC-GM medium to the contents of the vial dropwise with gentle agitation.
4. Transfer the contents immediately into a blue polypropylene 15-ml conical tube with 10–15 ml warm hESC-GM medium and centrifuge at 1000 rpm (210 g) for 4 min.
5. Aspirate the supernatant, add 1 ml hESC-GM, and gently resuspend the cells using a 1-ml pipetman with 2–4 repetitions.
6. Transfer the cells to the prepared PMEF plates with equilibrated hESC-GM medium. Spread the cells evenly throughout the well by moving the plate several times in two directions, at 90° to each other, and avoid swirling.
7. Check the cells the next day. If there are many dead cells or the medium has changed color, change 2/3 of the medium. Otherwise, do not change it for another day. On the second day after passage, change half (1.5 ml) of the medium every 24 h until the cells reach 70–80 % confluence (Fig. 1.1a).
8. Rinse the plate of hESCs with 2 ml Ca^{2+}, Mg^{2+}-free PBS 2–3 times and add 1 ml 0.05 % trypsin/0.53 mM EDTA to 1 well of 6-well plate.
9. Incubate at RT for 2–3 min and then pipette with a 1-ml pipetman to produce smaller cell clumps (2–5 cells).
10. Collect cells by adding 2 ml of PMEF medium and spin down at 1000 rpm (210 g) for 4 min.
11. Resuspend cells in 3 ml of hESC-GM and replate in 3 wells of 6-well plate with preformed PMEF feeder equilibrated with 2 ml hESC-GM.
12. Change half (1.5 ml) of the medium every 24 h until the cells reach 70–80 % confluence.

Fig. 1.1 Hemangioblast
development from hESCs
maintained on MEFs.
a hESCs on MEFs (200×);
b EBs from hESCs at day 4
(40×); and **c** Hemangioblast
colonies at day 6 (100×)

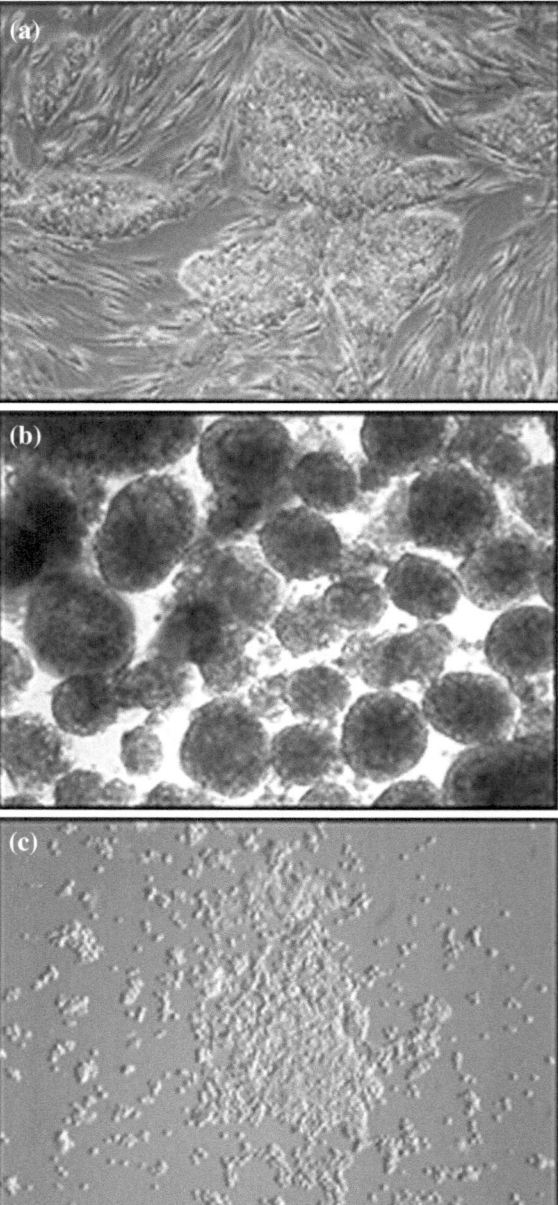

1.2.2.3 Generation of Hemangioblasts (Blast Cells) from hESCs [17, 18]

1. Collect undifferentiated hESCs by trypsinization. Usually, 1 well of 80 % confluent, high-quality undifferentiated hESCs (Fig. 1.1a) will generate approximately 2 million cells.
2. Plate cells in EB formation medium-I at a density of 2.5–5.0 × 10^5 cells/ml using Costar Ultra-low 24-well plate and incubate for 48 h. EBs form during the first 24 h.
3. Remove half (0.5 ml) of the medium with a 1-ml pipetman after 48 h and add 0.5 ml of EB formation medium-II without disturbing the EBs. Continue incubation for another 36 h (Fig. 1.1b).
4. Transfer EBs after 80–946 h of culturing (total EB formation time) into a 15-ml conical tube. Let stand for 1 min and aspirate medium gently.
5. Add 0.5 ml trypsin/EDTA and mix gently. Incubate 37 °C for 2–5 min.
6. Pipette vigorously with a 1-ml pipetman to dissociate EBs. If visible clumps still remain, incubate another 1–2 min and repeat pipetting as above until no visible clumps can be seen.
7. Add 2 ml serum-containing medium such as PMEF medium and pass through a 22G needle three times.
8. Count cells; usually 1 well of high-quality hES cells should generate 1.5–2 million EB cells. If yield is low, efficiency may not be good.
9. Spin down cells at 1000 rpm (210 g) for 4–5 min and resuspend cells in Stemline II hematopoietic stem cell expansion medium at a density of 2–5 × 10^6 cells/ml.
10. Mix 1.0–1.5 × 10^5 cells (<0.1 ml) with 2.5–3.0 ml of BGM. Votex for 10 s and let stand for 5 min.
11. Transfer the BGM cell mixture to 1 well of 6-well Costar Ultra-low plate by using a 3-ml syringe attached to a 16G needle and incubate at 37 °C with 5 % CO_2.
12. Check blast colony growth after 4 days. Usually, blast colonies are visible at 3 days, and after 4–6 days, grapelike blast colonies can be easily identified under microscopy (Fig. 1.1c). After 6–7 days, large, grape-like blast colonies can be picked up using a mouth-glass capillary tube or a P10 pipette tip for in vitro lineage differentiation studies and for in vivo functional studies.

1.3   Hemangioblast Generation from hESCs and iPSCs Cultured Under 3D Condition [19]

1.3.1   Materials

1. BD hESC-qualified Matrigel (BD, Cat#. 354277). Prepare aliquots according to the manufacturer's instruction. Aliquots may be stored at −80 °C for up to 6 months. The volume of the aliquots is typically between 270 and 350 μL;
2. mTeSR1 medium: add supplement (100 ml) to 400 ml basal medium and aliquots to 50–100 ml, stored at −20 °C. mTeSR1 medium can be stored at 4 °C for 2–3 weeks.
3. Matrigel coating: thaw aliquot (kept at −80 °C) overnight (ideally; or 3 h) at 4 °C; chill plates, glass pipet, and 12.5 ml DMEM/F12 in freezer before use; using 1-ml pipetman, add 1 ml of chilled DMEM/F12 medium to Matrigel aliquot, and pipet up and down to mix. Transfer to 15-ml Falcon tube with remaining DMEM/F12 medium and mix several times by inverting. Add 1 ml/well for 6-well plate and swirl the plate to spread the Matrigel solution evenly across the surface. For immediate use, keep at room temperature for at least 1 h before use. Do not remove Matrigel solution until the plates are ready to be used. For future use: Parafilm plate to prevent dehydration and store at 4 °C for up to 7 days, keep at room temperature for 0.5 h before use.
4. Preparation of microcarriers: siliconize glass bottle with Sylon CT (Sigma, Cat# 33065-U) 5 % DMDCS in toluene: swirl 5 ml around bottle surface in fume hood, evaporate overnight, and autoclave bottle. Weigh 5 g microcarriers: Whatman DE53 preswollen microgranular DEAE cellulose: (Whatman, Cat# 4058-050); transfer cellulose microcarriers to glass bottle add 100 ml PBS to microcarriers (Carrier concentration = 50 mg/ml) and adjust pH solution to 7.2 using 4 M HCl; let solution sit for 30 min, so that microcarriers settle to bottom of bottle, aspirate as much PBS as possible; bring volume of PBS to 100 ml and repeat steps 4–6 two times. After microcarriers have been adjusted for the third time, autoclave in glass bottle and store at 4 °C.
5. Coating microcarriers with Matrigel: pipet 3 ml of microcarrier solution (150 mg total carriers), centrifuge solution 2000 rpm for 4 min, and aspirate PBS supernatant; resuspend in 29 ml ice-cold mTeSR 1 medium in 50-ml tube and add 1 ml ice-cold Matrigel to mixture; nutate mixture overnight at 4 °C.

1.3.2   Methods

1.3.2.1   Two-Dimensional Growth and Expansion of hESCs and iPSCs

1. Gently tilt the plate onto one corner and allow the excess Matrigel solution to collect in that corner; remove the solution and add 2 ml mTeSR1 medium.

2. Quickly thaw hESCS or iPSCs in a 37 °C water bath by gently shaking the cryovial continuously until only a small frozen pellet remains, and remove the cryovial from the water bath and wipe with 70 % ethanol to sterilize.

3. Use a 2-ml pipette to transfer the contents of the cryovial to a 15-ml conical tube; add 5–7 ml of warm mTeSR1 dropwise to the tube and gently mix as the medium is added.

4. Centrifuge cells at 1000 rpm for 4 min at room temperature and gently aspirate the supernatant and keep the pellet intact.

5. Gently resuspend pellets in 2–3 ml mTeSR1 medium and avoid further breaking down cell clumps.

6. Transfer 1 ml of cell suspension to 1 well of BD Matrigel-coated 6-well plate. If one vial of the cells frozen from one 80–90 % confluence well, the cells can be thawed onto 2–3 well of Matrigel-coated 6-well plate.

7. Place the plate into the 37 °C incubator with 5 % CO_2 and 95 % humidity, and move the plate in quick side-to-side, forward-to-back motions to evenly distribute the clumps within the wells.

8. Perform daily medium changes. At 48 h post-thaw, change media 3 ml/well and then change media daily until confluent.

9. When confluent, split cells 1:6. Rinse wells 1 time with 2 ml DMEM/F12 per well and aspirate medium; add 1 ml dispase (1 mg/ml) solution to each well and incubate 8 min at 37 °C; rinse well 3 times with 2 ml DMEM/F12 per well, aspirate medium, add 1 ml mTeSR1 medium per well, and remove cells from surface by scraping with pipet tip.

10. Collect cell suspension, rinse each well with 1 ml mTeSR1 medium, and pool cells; add mTeSR medium to cell suspension to make total volume = 20 ml and add 10 µl of Y-27632 ROCK inhibitor to medium to make final concentration = 5 µg/ml.

11. Plate 3 ml of cell suspension per well with Matrigel-coated plate and incubate at 37 °C as above.

1.3.2.2 Culturing hESCs and iPSCs on 3D Microcarriers

1. Place Matrigel-coated microcarriers at room temperature for 1 h.

2. When cells are confluent, dissociate cells using dispase as described above and collect cells by centrifugation at 1000 rpm for 5 min.

3. Aspirate supernatant, add 3 ml Matrigel-coated microcarriers/well of cells, and seed cells in 1 well/6-well ultra-low plate.

4. Place plate on shaker at 37 °C, shake for 2 h ∼ 100 rpm, and then incubate under static conditions.

5. For first and second media changes, first remove cell suspension from tissue culture dish using 5-ml pipet, transfer to 15-ml tube, and let cell suspension sit for 15 min; after this time, all cells adhered to microcarriers will settle at bottom of tube.

6. Gently aspirate supernatant, taking care not to aspirate cell pellet, resuspend in 4 ml mTeSR 1 medium, and reseed cells back into tissue culture dish with 5-ml pipet.
7. After 2 media changes, media can be changed either by the method described above or by tilting the tissue culture dish at an angle and aspirating medium away from cells. Be careful not to aspirate cell clumps on microcarriers. This method requires more intricate skill and larger cell aggregates to be in culture (Fig. 1.2).

1.3.2.3 Passaging Cells Grown on Microcarriers

1. Collect cells with 5-ml pipette, transfer to 15-ml conical tube, and let cells sit in tube ~1–2 min and aspirate medium.
2. Enzymatic passaging of hESCs/hiPSCs on microcarriers: add 2 ml dispase to microcarrier + hESC/hiPSCs and incubate at 37° for 8 min; pipette up and down to break cell clumps from microcarriers. Rinse with DMEM/F12 twice to remove any extra dispase. Pass cells through a 70–100-μm strainer to remove excess microcarriers from culture. Collect cells by centrifugation; resuspend cells in 15 ml Matrigel-coated microcarriers and plate in 5 wells/6-well plate. Place plate on shaker at 37 °C, shake for 2 h ~100 rpm, and then incubate under static conditions.
3. Mechanical passaging of hESCs/hiPSCs on microcarriers: add 2 ml of fresh medium to microcarrier–cell pellet; pipette up and down with a P1000 pipet tip to break cells from microcarriers; and pass through a 70–100 μm strainer to remove excess microcarriers from culture. Plate cells as above.

1.3.2.4 Embryoid Body Formation and Hemangioblast Generation

1. Collect cells with 5-ml pipette, transfer to 15-ml conical tube, and let cells sit in tube ~1–2 min and aspirate medium.
2. Resuspend with 5 ml PBS and centrifuge 1000 rpm 1 min (or let cells sit 2 min in tube), and aspirate supernatant.
3. Resuspend in 3 ml EB medium-I. On day 2, replace half of medium with fresh EB medium-II.
4. Collect EBs (Fig. 1.3a, b) on day 3.5–4.0, trypsin to obtain single cell suspension, and plate for hemangioblast formation as described in Sect. 1.2.2.3 (Fig. 1.3c).

Fig. 1.2 Expansion of hESCs on microcarriers: hESCs were cultured on two-dimensional Matrigel with mTeSR1 medium, and then collected and replated on three-dimensional microcarriers for 4 h (A, ×200), 2 days (B, ×100), and 4 (C, ×100) days. Robust expansion of hESCs was observed for 2 days after transfer from two-dimensional Matrigel to three-dimensional microcarrier condition. (This figure was originally published in Regenerative Medicine. Lu et al. [19], **with permission from Future Medicine**)

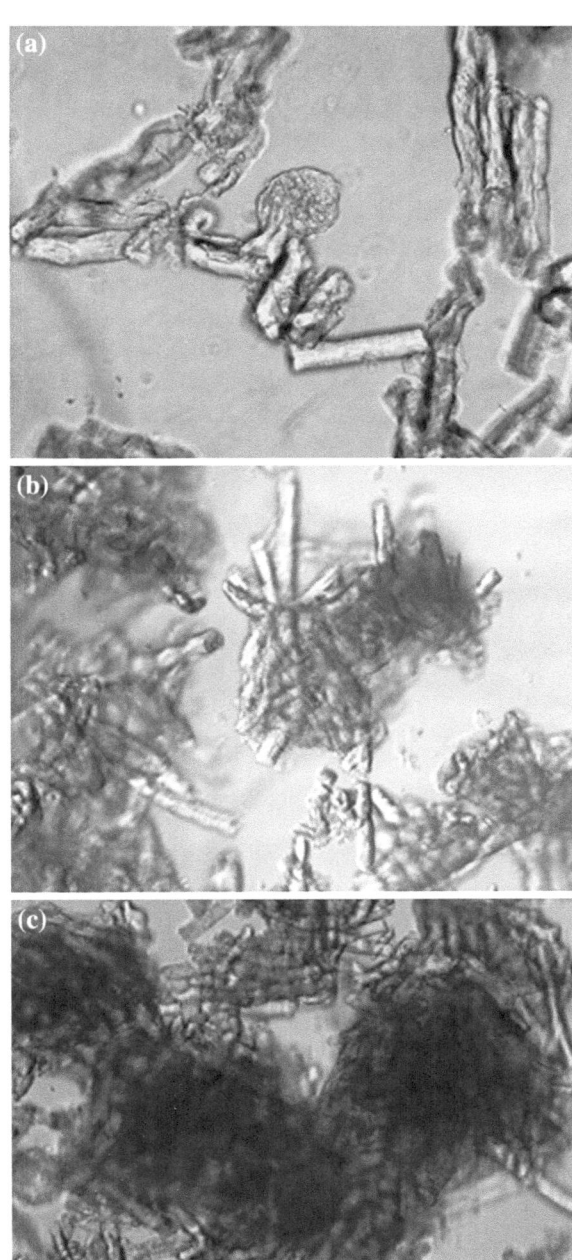

Fig. 1.3 EB formation and hemangioblast development of hESCs and iPSCs maintained on microcarriers. hESCs and iPSCs were cultured on microcarriers for 4 days and then switched to EB formation medium for another 4 days. **a** EBs from hESCs (100×) and EBs from iPSCs (40×) were collected and trypsined, and single cells were plated for hemangioblast development for 6 days (C, 100×)

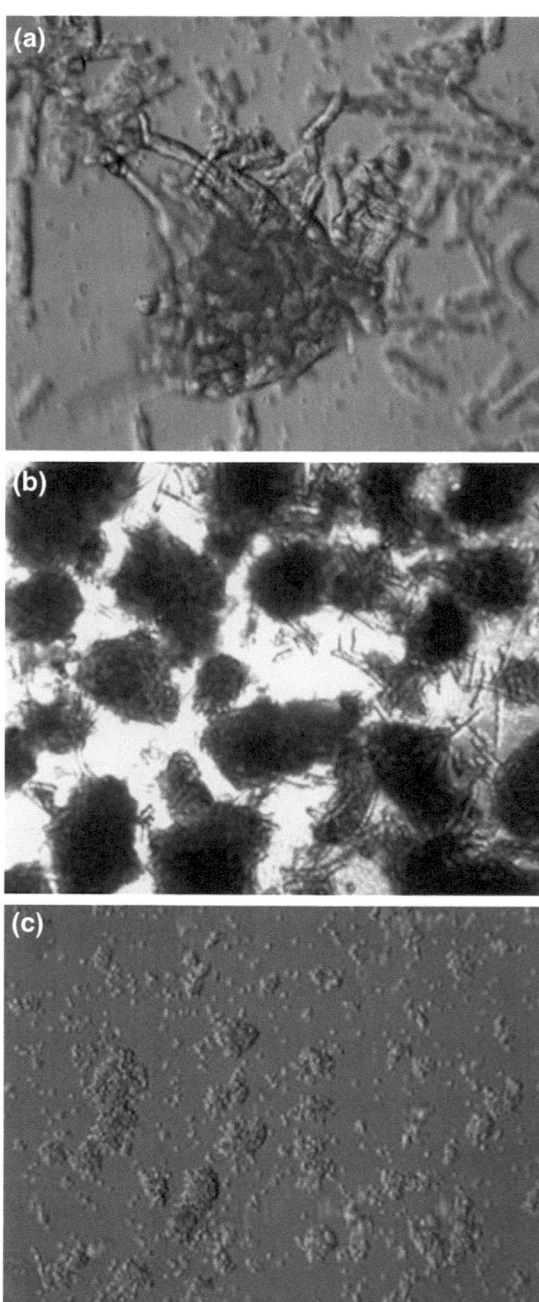

References

1. Wagner RC. Endothelial cell embryology and growth. Adv Microcirc. 1980;9:45–75.
2. Ema M, Faloon P, Zhang WJ, et al. Combinatorial effects of Flk1 and Tal1 on vascular and hematopoietic development in the mouse. Genes Dev. 2003;17:380–93.
3. DSouza SL, Elefanty AG, Keller G. SCL/Tal-1 is essential for hematopoietic commitment of the hemangioblast but not for its development. Blood 2005;105:3862–3870.
4. Lu SJ, Ivanova Y, Feng Q, Luo C, Lanza R. Hemangioblasts from human embryonic stem cells generate multilayered blood vessels with functional smooth muscle cells. Regen Med. 2009;4:37–47.
5. Wang X, Kimbrel EA, Ijichi K, et al. Human ESC-derived MSCs outperform bone marrow MSCs in the treatment of an EAE model of multiple sclerosis. Stem Cell Reports. 2014;3:115–30.
6. Asahara T, Murohara T, Sullivan A, et al. Isolation of putative progenitor endothelial cells for angiogenesis. Science. 1997;275:964–7.
7. Murohara T, Ikeda H, Duan J, et al. Transplanted cord blood-derived endothelial precursor cells augment postnatal neovascularization. J Clin Invest. 2000;105:1527–36.
8. Hungerford JE, Little CD. Developmental biology of the vascular smooth muscle cell: building a multilayered vessel wall. J Vasc Res. 1999;36:2–27.
9. Conway EM, Collen D, Carmeliet P. Molecular mechanisms of blood vessel growth. Cardiovasc Res. 2001;49:507–21.
10. Biomedicine Carmeliet P. Clotting factors build blood vessels. Science. 2001;293:1602–4.
11. Choi K, Kennedy M, Kazarov A, Papadimitriou JC, Keller G. A common precursor for hematopoietic and endothelial cells. Development. 1998;125:725–32.
12. Kennedy M, Firpo M, Choi K, et al. A common precursor for primitive erythropoiesis and definitive haematopoiesis. Nature. 1997;386:488–93.
13. Wang L, Li L, Shojaei F, et al. Endothelial and hematopoietic cell fate of human embryonic stem cells originates from primitive endothelium with hemangioblastic properties. Immunity. 2004;21:31–41.
14. Zambidis ET, Peault B, Park TS, Bunz F, Civin CI. Hematopoietic differentiation of human embryonic stem cells progresses through sequential hematoendothelial, primitive, and definitive stages resembling human yolk sac development. Blood. 2005;106:860–70.
15. Umeda K, Heike T, Yoshimoto M, et al. Identification and characterization of hemoangiogenic progenitors during cynomolgus monkey embryonic stem cell differentiation. Stem Cells. 2006;24:1348–58.
16. Kennedy M, D'Souza SL, Lynch-Kattman M, Schwantz S, Keller G. Development of the hemangioblast defines the onset of hematopoiesis in human ES cell differentiation cultures. Blood. 2007;109:2679–87.
17. Lu SJ, Feng Q, Caballero S, et al. Generation of functional hemangioblasts from human embryonic stem cells. Nat Methods. 2007;4:501–9.
18. Lu SJ, Luo C, Holton K, et al. Robust generation of hemangioblastic progenitors from human embryonic stem cells. Regen. Med. 2008;3:693–704.
19. Lu SJ, Kelley T, Feng Q, et al. 3D microcarrier system for efficient differentiation of human pluripotent stem cells into hematopoietic cells without feeders and serum [corrected]. Regen Med. 2013;8:413–24.
20. Oh SK, Chen AK, Mok Y, et al. Long-term microcarrier suspension cultures of human embryonic stem cells. Stem Cell Res. 2009;2:219–30.
21. Klimanskaya I, Chung Y, Becker S, Lu SJ, Lanza R. Derivation of human embryonic stem cells from single blastomeres. Nat Protoc. 2007;2:1963–72.
22. Klimanskaya I, McMahon J. Approaches of derivation and maintenance of human ES cells: detailed procedures and alternatives. In: Lanza Rea, ed. Handbook of Stem Cells. Volume 1: Embryonic Stem Cells. New York, USA: Elsevier/Academic Press; 2004:437–449.

Chapter 2
Derivation of Mature Erythrocytes from Human Pluripotent Stem Cells by Coculture with Murine Fetal Stromal Cells

Bin Mao, Xulin Lu, Shu Huang, Jinfeng Yu, Mowen Lai, Kohichiro Tsuji, Tatsutoshi Nakahata and Feng Ma

Abstract Transfusion of red blood cells (RBCs) is a requisite cell therapy today, while RBCs supplied by donors cannot match the huge demand of patients. Human pluripotent stem cells (hPSCs) are promising cell sources to obtain RBCs as an alternative transfusion product for clinical application. Several in vitro culture systems have been reported that in which mature erythrocytes can be efficiently generated from hPSCs. However, different efficiency and maturity of hPSC-derived erythrocytes could be obtained when using different culture systems. We still lack a complete understanding of the regulatory pathways controlling human erythrocyte development and maturation, especially the origination of erythrocytes early in the embryo and enucleation at the terminal stage of differentiation. In this chapter, we focus on an efficient method established successfully in our laboratory to derive functionally mature erythrocytes from hPSCs by coculture with mouse fetal stromal cells [aorta–gonad–mesonephros stromal cells (mAGM) and fetal liver stromal cells (mFLSCs), respectively]. The procedures to investigate the characteristics of these hPSC-derived erythrocytes are also introduced, including colony formation assay to detect the hematopoietic potential, flow cytometry assay to detect the phenotypic

B. Mao · X. Lu · S. Huang · J. Yu · M. Lai · F. Ma (✉)
Institute of Blood Transfusion, Chinese Academy of Medical Sciences and Peking Union Medical College, 26 Huacai Road, Longtan Industry Park, Chenghua, 610052 Chengdu, China
e-mail: mafeng@hotmail.co.jp

K. Tsuji · F. Ma
Institute of Medical Science, The University of Tokyo, Tokyo, Japan

T. Nakahata
Center for IPS Cell Research and Application (CiRA), Kyoto University, Kyoto, Japan

F. Ma
State Key Lab of Experimental Hematology, Chinese Academy of Medical Sciences and Peking Union Medical College, Tianjin, China

© The Author(s) 2015
T. Cheng (ed.), *Hematopoietic Differentiation of Human Pluripotent Stem Cells*,
SpringerBriefs in Stem Cells, DOI 10.1007/978-94-017-7312-6_2

expression pattern, and immuno-staining assay of the Hb components to evaluate the maturity. At the end of this review, several future prospects are also be addressed in this research fields.

Keywords Erythrocytes · hPSCs · AGM · Fetal liver · Hematopoiesis

2.1 Introduction

Transfusion of red blood cells (RBCs) is an indispensable procedure used in the clinic today. There is a notable imbalance between the demand of patients and the supply of donors, especially in developing countries [1]. The safety and sufficiency of the blood supply are national and international priorities [2]. Immense efforts have been made to enhance the supply of RBCs in vitro as an alternative transfusion product since two decades ago. Studies pioneered by L Douay's group prove that it is possible to produce RBCs from human adult hematopoietic stem cells/hematopoietic progenitor cells (HSCs/HPCs) of cord blood (CB), bone marrow (BM), and peripheral blood (PB) [3–5]. This group reported that the autologous cultured RBCs (cRBCs) derived from peripheral CD34$^+$ cells were transfused back into the human recipient. The survival of cRBCs was successful as long as natural counterparts do through detecting the 51Cr labeled on cRBCs [6]. This work set up an example of transfusion cRBCs generated from adult HSCs/HPCs in vitro for the first time. However, so far it is still a thorny problem for us to expand adult HSCs/HPCs in vitro [7, 8], and the productivity of RBCs is related to cell sources from different individuals [9]. Thus, to a great extent, the progression of using adult HSCs/HPCs to produce large scale of RBCs for transfusion purpose is hampered.

Human pluripotent stem cells (hPSCs), including human embryonic stem cell (hESCs) [10] and induced pluripotent stem cells (iPSCs) [11], have the ability of unlimited self-renewal and pluripotency. By specific inducing method, large number of hematopoietic cells can be generated from these hPSCs, indicating they might be a potential cell sources to obtain large-scale RBCs for clinical application. cRBCs of specific blood group, rare blood group, or null blood group could be produced from hPSCs by genetically modified; thus, it may satisfy an ideal transfusion for those who with rare blood types. In addition, induced pure cRBCs without other granulocytes and lymphocytes would reduce transfusion reactions and graft-versus-host disease (GVHD) in immunocompromised patients.

Besides as cell source for clinical application, hPSC-derived erythrocytes also offer a subtle model to study human erythropoiesis in vitro. Previously, the mechanism controlling early development of human embryonic/fetal hematopoietic is largely unknown, because there lacks a proper experimental model to mimic the early progress in human ontogeny. Erythrocytes generated from hESCs support an excellent platform to study on the origination, development, and maturation of erythroid cells derived from the earliest endothelial/hematopoietic progenitors and to uncover important regulating mechanisms at each developmental stage step by step. Since hiPSCs can be established from specific individual patients, they also provide a powerful tool for modeling diseases and supply a new opportunity for patient-tailored therapies. Several reports have referred to the successful establishment of hiPSC lines derived from patients with erythrocytes-associated diseases [12–17].

Schemes of successfully inducing large-scale production of erythrocytes from hPSCs have been reported by several groups, including our laboratory. Generally, hematopoietic progenitors are produced first through the formation of embryonic bodies (EBs) spontaneously by hPSCs [17–23], or coculture with stromal cells isolated from hematopoietic niches [24–29]. Then, hPSC-derived multipotential hematopoietic progenitors are differentiated into erythrocytes directionally (and other blood lineage cells, as well). At last, erythrocytes undergo maturation with or without feeder cells. Based on the establishment of methods to induce erythrocytes from hPSCs in vitro, more and more information relating development of human early erythropoiesis has been accumulated. Yet we are still lacking of a complete understanding of the regulatory pathways controlling erythrocyte development and maturation, especially the origination and initiation of erythrocytes early in the embryo and the enucleation at the terminal stage of differentiation. It has long been known that there are two erythropoietic waves in human. The first wave of primitive erythrocytes, which are originated from the extra-embryonic mesoderm of yolk sac, synthesize Hb Gower I ($\zeta_2\varepsilon_2$) and Hb Gower II ($\alpha_2\varepsilon_2$) [30]. The second wave of definitive erythrocytes, which are originated from the AGM region of the embryo proper [31], express ζ-, ε-, α-, γ-, and small amount of β-globin just at the beginning, and then, ζ- and ε-globins are silenced rapidly. Around birth, BM becomes the main site of erythropoiesis and expresses β-globins which almost replaces γ-globin expression [25]. So the hemoglobin (Hb) contents can be used to evaluate the maturity of erythrocytes. Consistent data of Hb contents from different laboratories demonstrate that hPSC-derived erythrocytes from coculture system, which express higher adult β-globin, are more mature than erythrocytes generated from EBs (Table 2.1). Affording appropriate microenvironment is indispensable for the maturation of erythrocytes. The common stromal cells used for stimulating the generation of hematopoietic cells from hPSCs include mouse BM stromal cell lines: OP9, MS-5, S17, mAGM, mFLSCs, FH-B-hTERT (immortalized human fetal liver hepatocyte line), and so on. Although the OP9 cell line, which is used most frequently, has a good ability to induce the derivation of early hematopoietic cells, it is obtained from gene mutation mouse which cannot produce macrophage colony-stimulating factor (M-CSF). So the hematopoiesis induced by coculture with OP9 could not stand for the natural development process.

We have established an efficient system by coculture hPSCs with mAGM cell lines [32, 33] and mFLSCs [26, 27] (Fig. 2.1), which were obtained from normal fetal mice, to generate robust growth of HPCs [26]. When coculture with these early definitive hematopoiesis-supporting stromal cells, considerable hPSC-derived E colonies and E bursts could be generated on day 10–14 following hematopoietic

Table 2.1 Erythrocytes derived from hPSCs in vitro by different methods

	Methods	Expansion folds	Expression ratio of Hb-β	Enucleation ratio	Ref.
EB	Step 1: hESC→EBs, SF/SC	High (10^3-fold)	Little (2 %) or no	High (34–48 %)	[20, 22]
	Step 2: differentiation to erythrocyte, SF				
	Step 1: hiPSCs→EBs, SF	Moderate (10^2-fold)	No	Low (4–10 %)	[20, 23]
	Step 2: differentiation to erythrocyte, SF				
EB + coculture	Step 1: hESC→EBs, SF		Low (0–16 %)	High (30–65 %)	[18]
	Step 2: differentiation to erythrocyte, SF				
	Step 3: enhancing enucleation ratio by coculture with stromal cells (OP-9/MS-5), SF				
Coculture	Step 1: hESC + stromal cells, SC	Moderate (10^2-fold)	High (99.8 ± 0.6 %)		[24, 25, 27, 28]
	Step 2: expansion of hemangioblasts, SC				
	Step 3: differentiation to erythrocyte, SC/SF				
	Step 4: enhancing enucleation ratio by coculture with stromal cells (OP-9/MS-5), SF				

SC Serum-containing culture; *SF* Serum-free culture

colony culture. By a clone tracing method, we demonstrate that most hESC-derived erythroid colonies expressed adult β-globin and gradually increased to almost 100 % with additional 6 days liquid culture. In addition, the hESC-derived erythrocytes can undergo enucleation and carry and release oxygen functionally [27] (Fig. 2.2). These data indicate that the coculture of hPSCs with definitive hematopoiesis-niche-derived stromal cells is an efficient method to develop hematopoiesis system and generate adult-type erythrocytes with functional maturity. Using this model, we recently have observed the earliest erythrocytes development from hPSCs, which emerge earlier than the definitive hematopoiesis takes place. The phenotypic pattern of hPSC-derived early erythrocytes is distinct from erythrocytes derived from adult HSCs/HPCs. Since there is some difficulty to gain large-scale production of murine fetal liver-derived stromal lines [34], the system of hPSCs coculture with mAGM is superior for looking into the origination of the earliest erythroid cells.

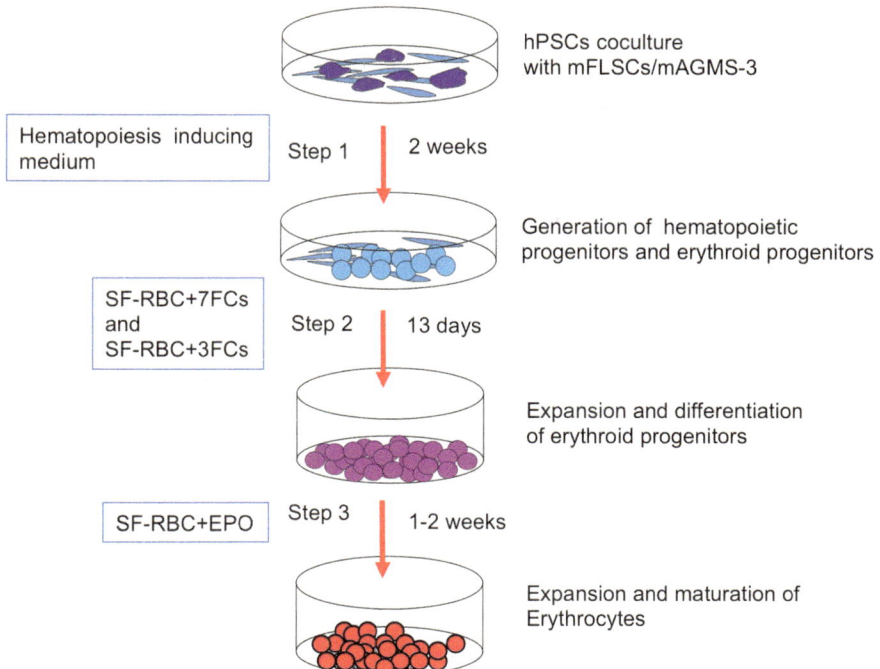

Fig. 2.1 The illustration of the induction method of erythrocytes derived from hPSCs in our laboratory. The induction procedure of erythrocytes differentiation was divided into three steps. *Step 1* Undifferentiated hPSC colonies were individually picked up from the primary culture manually (routinely $0.5–1 \times 10^3$ cells/colony). Coculture the colonies with irradiated mFLSCs/mAGMS-3 about 2 weeks to generate multipotential hematopoietic progenitors and erythroid progenitors. *Step 2* To expand and differentiate the erythroid progenitors, treat the cocultures with 0.25 % Trypsin/EDTA and transfer all the harvested cells into suspension culture medium. First with a cocktail of SF-RBC + 7FCs (IMDM + SCF, VEGF, TPO, FL, IL-6, IL-3, EPO) for about 6 days and then in SF-RBC + 3FCs (IMDM + SCF, IL-3, EPO, dexamethasone) for about 7 days. *Step 3* The erythrocytes develop and mature further in the SF-RBC + EPO medium but without any other cytokines

In this chapter, we here introduce a coculture system developed in our laboratory to produce erythrocytes efficiently from hPSCs. Since the derivation of mAGMS and mFLSCs has been described elsewhere [32, 34], we will focus on the methodological processes of generating large number of functionally mature erythrocytes. Our method include three steps: (1) generation of erythroid cells and other hematopoietic cells from cocultures; (2) expansion of erythroid progenitors derived from cocultures in suspension culture; and (3) further maturation of erythrocytes (Fig. 2.1). We also introduce the procedures for assaying the erythroid (and other hematopoietic lineages) differentiation in semisolid culture, detecting phenotypic expression pattern by flow cytometry assay and evaluating the maturity level of derived erythrocytes by Hb immuno-staining assay, respectively. And the pivotal

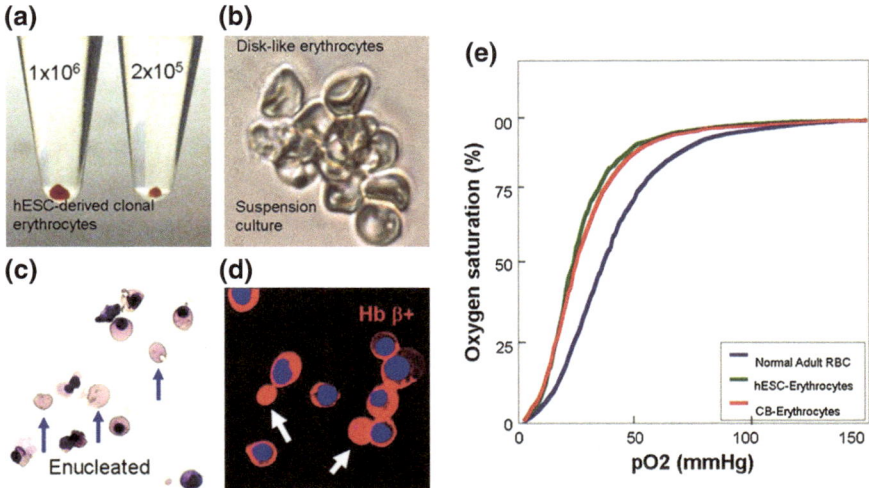

Fig. 2.2 Functionally mature erythrocytes derived from hESCs/mFLSCs coculture. **a** Photograph of harvested large BFU-E colony cells from day 16 coculture, showing the *red color* of human erythrocytes. A total of 2×10^5 (*right*) and 1×10^6 (*left*) erythroid cells were collected from one and five large BFU-E colonies, respectively. **b** Cluster of enucleated erythrocytes derived from hESCs from the same suspension cultures shown in (**c**) and (**d**). **c** Sample of hESC-derived erythroid cells from a day 12 + 6 suspension culture (May-Grünwald-Giemsa staining, MGG). *Arrows* indicate enucleated erythrocytes. **d** Immuno-staining for β-globin expression in hESC-derived erythroid cells. *Arrows* indicate β-globin expressing enucleated erythrocytes. **e** Oxygen dissociation curves of day 15 hESC/mFLSC coculture-derived clonal large BFU-E erythroid cells at day 16 of the colony culture compared with hCB-derived BFU-E cells and normal peripheral blood cells. Adapted from Ref. [27]

details during the experimental operation that should pay attention to are described additionally as notes in this protocol. At the end of this chapter, the hampers existed in the process of clinical translation of hPSC-derived erythrocytes and the methodological advantage of our coculture system to resolve these puzzles are also addressed.

2.2 Materials

2.2.1 hPSCs Lines

1. hESC line, H1, was obtained from WiCell Research Institute (Madison, WI), and KhES-3 was kindly provided by Professor N Nakatsuji at Department of Development and Differentiation, Institute for Frontier Medical Sciences, Kyoto University, Japan.

2. hiPSC 201B7 line was kindly provided by Professor S Yamanaka at Center for iPS Cell Research and Application, Kyoto University, Japan.
3. hESC-/iPSCs-maintaining medium.

   - Dulbecco's modification of Eagle's medium (DMEM; Gibco, Cat. No. 12800-017),
   - F12 (Gibco, Cat. No. 21700-075),
   - KnockOut™ Serum Replacement (KSR; Gibco, Cat. No. 10828-028),
   - 2-mercaptoethanol (2-ME; Sigma, Cat. No. M7522),
   - L-glutamine (Gibco, Cat. No. 25030-081),
   - Nonessential amino acid solution (NEAA; Gibco, Cat. No. 11140-050), and
   - Basic fibroblast growth factor (bFGF; Gibco, Cat. No. 13256-029).

4. Gelatin-coated 10-cm culture dish (Corning, Cat. No. 430167).
5. Scraper (Nunc, Cat. No. 179693).

2.2.2 mFLSCs (Adapted from Refs. [27, 34])

1. Mice (Pregnant day 14–15, Strain: C57/Black6).
2. Ophthalmology surgery scissors and forceps (Autoclaved before use).
3. Sterile tissue culture dish (Corning, Cat. No. 3296).
4. Gelatin-coated 6-well culture plate (Corning, Costar, Cat. No. 3335).
5. Dulbecco's phosphate-buffered saline without Ca^{2+} and Mg^{2+} (D-PBS (-); Wako, Cat. No. 045-29795).
6. Trypsin/EDTA solution [Gibco, Cat. No. 25200-056/0.25 %, diluted to 0.05 % with D-PBS (-)].
7. mFLSCs culture medium

   - DMEM medium (Gibco, REF. 31600-034).
   - Fetal bovine serum (FBS; Biowest, Cat. No. S1580-500) 10 % in volume.

2.2.3 mAGMS-3 Cell Line Culture (Adapted from Ref. [32])

mAGMS-3 cell line derived from murine AGM region had been established and maintained in our laboratory since 1998 [32] with stable and efficient hematopoiesis-supporting potential.

1. mAGMS-3 cell line (between passage 30 (P 30) and P 40).
2. Gelatin-coated 10-cm culture dish (Corning, Cat. No. 430167).

3. T225 flask (Nunc. Cat. No. 159934).
4. Cryogenic vial (Nunc. Cat. No. 377267).
5. mAGMS-3 cell-maintaining medium:

   - Minimum essential medium, alpha modified (α-MEM; Hyclone, Cat. No. SH30265) and
   - FBS (Biowest, Cat. No. S1580-500) 5 % in volume.

2.2.4   Induction of Multipotential Hematopoietic Progenitors

1. Biological X-ray irradiator (Rad Source Technologies, Inc. RS2000).
2. Gelatin-coated 6-well culture plate (Corning, Cat. No. 3335).
3. Undifferentiated hPSC colonies.
4. Radiated mFLSCs or mAGMS-3 cells (Radiation dose: 25 Gy for mFLSCs and 15 Gy for mAGMS-3 cells).
5. Hematopoiesis-inducing medium in coculture:

   - Iscove's modified Dulbecco's medium (IMDM; Gibco, Cat. No. 12440-053),
   - FBS (Hyclone; 10 % in volume),
   - NEAA (Gibco, Cat. No. 11140-050),
   - Ascorbic acid (AA; Sigma, Cat. No. 1043003-1G),
   - Transferrin (Sigma, Cat. No. T2252),
   - recombinant human Vascular endothelial growth factor (rhVEGF; WAKO, Cat. No. 229-01353), and
   - 2-ME (Sigma, Cat. No. M7522).

6. Y-27632 (CALBOCHEM, Cat. No. 68800).

2.2.5   Suspension Culture in Liquid Medium

1. Erythroid progenitors expansion medium, named as SF-RBC + 7FCs medium:

   - Serum-free expansion medium (SFEM; Stemcell, Cat. No. 09650),
   - Bovine serum albumin (BSA; Sigma, Cat. No. A-4161),
   - Transferrin (Sigma, Cat. No. T2252),
   - 2-ME (Sigma, Cat. No. M7522),
   - rhVEGF (Wako, Cat. No. 229-01353),
   - rhStem cell factor (rhSCF, Wako, Cat. No. 199-12813),
   - rhInterleukin 3 (rhIL-3, Kirin, Cat. No. MYE0317),

- rhInterleukin 6 (rhIL-6, Wako, Cat. No. 099-04631),
- rhFlt-3 ligand (rhFL, Wako, Cat. No. 061-04051),
- rhTrombopoietin (rhTPO, Kirin, Cat. No. NHK0823-SDM), and
- rhErythropoietin (rhEPO, provided by Kirin Brewery Company, Tokyo, Japan).

2. Erythrocytes expansion medium, named as SF-RBC + 3FCs medium:

- SFEM (Stemcell, Cat. No. 09650),
- BSA (Sigma, Cat. No. A-4161),
- Transferrin (Sigma, Cat. No. T2252),
- 2-ME (Sigma, Cat. No. M7522),
- rhSCF (Wako, Cat. No. 199-12813),
- rhIL-3 (Kirin, Cat. No. MYE0317),
- rhEPO (provided by Kirin Brewery Company, Tokyo, Japan), and
- Dexamethasone (Sigma, Cat. No. D4902).

3. Ultra-low adherent cluster 6-well plate (Corning, Cat. No. COSTAR® 3471).
4. hPSC-derived hematopoietic cells (after 0.25 % trypsin/EDTA treatment, total harvested cells were plated).

2.2.6 Hematopoietic Colony Culture

1. Semisolid culture medium:

- α-Methylcellulose (α-MTC; Sigma, Cat. No. M0512),
- α-MEM (Gibco, REF. 41061-029),
- FBS (Hyclone; 56 °C/30 min heat inactivated),
- BSA (Sigma, Cat. No. A-4161),
- 2-ME (Sigma, Cat. No. M7522),
- rhSCF (Wako, Cat. No. 199-12813),
- rhIL-3 (Kirin, Cat. No. MYE0317),
- rhIL-6 (Wako, Cat. No.099-04631),
- rhFL (Wako, Cat. No. 061-04051),
- rhTPO (Kirin, Cat. No. NHK0823-SDM),
- rhEPO (provided by Kirin Brewery Company, Tokyo, Japan), and
- rhGranulocyte colony-stimulating factor (rhG-CSF, Kirin, Cat. No. NSG0205-G604).

2. hPSC-derived hematopoietic cells (after 0.25 % trypsin/EDTA treatment, total harvested cells were plated).
3. 35-mm Petri dish (Becton Dickinson Labware, Cat. No. 35-1008).
4. 2-mL plastic syringe (Top Surgical Taiwan Corporation).
5. 18-gauge (18G) syringe needle (Terumo, Cat. No. 1838S).

2.2.7 Flow Cytometric Analysis of hPSC-Derived Erythrocytes

1. Trypsin/EDTA solution (Gibco, Cat. No. 25200-056/0.25 %).
2. Sorting medium (SM):

    - D-PBS (-) (Wako, Cat. No. 045-29795),
    - FBS (Biowest, Cat. No. S1580-500) 5 % in volume, and
    - Penicillin–streptomycin (PS; Hyclone, Cat. No. SV30010) 1 % in volume.

3. Antibodies:

    - Mouse antihuman c-kit, APC-conjugated (eBioscience, Cat. No. 17-1179-42),
    - Mouse antihuman glycophorin A (GPA) or CD235a, PE-conjugated (DAKO Cytomation, Cat. No. R 7078),
    - Mouse antihuman CD34, FITC-conjugated (BD Pharmingen™, Cat. No. 555821),
    - Mouse antihuman CD45, APC-conjugated (BD Pharmingen™, Cat. No. 559864),
    - Mouse antihuman CD47, FITC-conjugated (BD Pharmingen™, Cat. No. 556045),
    - Mouse antihuman CD71, FITC-conjugated (BD Pharmingen™, Cat. No. 555536),
    - Mouse antihuman CD81, FITC-conjugated (BD Pharmingen™, Cat. No. 551108),
    - Mouse antihuman erythropoietin receptor (EPO-R), FITC-conjugated (R&D Systems, Cat. No. FAB307F), and
    - 7-amino-actinomycin D (7-AAD; BD Pharmingen™, Cat. No. 559925).

4. Flow cytometry system (Becton Dickinson Company, FACSCanto™ II).
5. Rabbit serum (Zhongshan, ZLI9025).

2.2.8 Immuno-Staining Assay of hPSC-Derived Erythrocytes

1. Cells harvested on different days were cytospined on clean glass slides.
2. Antibodies (Abs):

    - Goat antihuman hemoglobin (Hb; Bethyl Laboratories, Cat. No. A80-134A),
    - Mouse antihuman Hbε (Santa Cruz Biotech, Cat. No. sc70421),
    - Mouse antihuman Hbγ (Santa Cruz Biotech, Cat. No. sc-21756),
    - Mouse antihuman Hbβ (Santa Cruz Biotech, Cat. No. sc-21757),

- Donkey antigoat Cy3-conjugated secondary Ab (Jackson Immuno Research, Cat. No.705-165-003), and
- Donkey antimouse FITC-conjugated secondary Ab (Jackson Immuno Research, Cat. No. 715-095-150).

3. Liquid marker pen (DAKO, Cat. No. 69932).
4. Skim milk (SM; BD, REF: 232100).
5. 4 % paraformaldehyde solution (4 %PFA; Boster, Cat. No. AR1608).
6. Triton X-100 (Uni-chem, Cat. No. 9002-93-1).
7. 4′,6-diamidino-2-phenylindole (DAPI; Roche, REF. 0 236 276 001).
8. D-PBS (-) (Wako, Cat. No. 045-29795).
9. Mounting medium (Vector Laboratories, Cat. No. H-100).
10. Microcover glass (Matsunami, 24 × 24 mm).
11. Nail polish.
12. Cell cytospin machine (Thermo Scientific, Cytospin 4).
13. Fluorescence microscope (Olympus, BX53).

2.3 Methods

In mammals, the hematopoietic system developed during early embryonic development starts in the yolk sac, transfers to intra-embryonic sites, initially to the aorta-gonad-mesonephros (AGM) region, and then processes to fetal liver until birth. Late in gestation, hematopoietic precursors settle in BM, where HSCs predominantly stay and recruit life long [35]. To our experiences, an efficient in vitro generation of multipotential HPCs from hPSCs is mostly depended on the supporting role of the mouse fetal stromal cells. Fetal liver is the dominant region for definitive hematopoiesis before birth. The coculture system of hPSCs with mFLSCs established by us can produce a large quantity of functional mature erythrocytes, and a progressively maturation process of hPSC-derived erythrocytes could be observed [26]. Because only primary mFLCs give rise to efficient hematopoiesis-induction effects in coculture and it is difficult to make large-scale expansion of efficient mFLSC lines, hPSCs coculture with mFLSCs shows some instability when used to gain large-scale experiments for blood cell production. Since the AGM region has been recognized as the earliest microenvironment to support the origination of definitive hematopoiesis [36–38], we then optimize our coculture system by using mAGM cell line as the hematopoiesis-inducing cells. The mAGM-S3 cell line was derived from the murine AGM region of a 10.5 days postcoitum (dpc) mouse embryo and reported previously [32]. In our hPSC-mAGM-S3 coculture system, robust growth of human early hematopoietic stem/progenitor cells could be achieved. By lineage-specific inducing method (such as toward erythroid development), we can harvest large quantity of functionally mature blood cells, including mature erythrocytes and innate immune-related cells (mast cells, eosinophils, basophilic cells, etc.). Consistent with our data, it has been reported that the mAGM stromal cells show more efficient hematopoietic supporting potential than mFLCs do [39]. Hence, the coculture system using mAGM cells provides a promising platform for us to gain both enough hPSC-derived mature erythrocytes and the understanding how the earliest erythroid cells come out.

2.3.1 Maintenance of hPSCs Lines

In short, the hPSCs lines (H1, H9, KhES-3, 201B7) can be maintained and passaged weekly on irradiated mouse embryonic fibroblast (MEF) feeder cells as described [10, 11].

Notes:

1. Prepare MEF cells well from 13.5 dpc mouse embryo. Routinely, MEF cells of passage3 (P3) to P5 are suitable for maintaining the undifferentiated hPSCs.
2. For maintaining the hPSCs in a long term, the physical method of scraping the colonies with pipette tips and then pipetting colonies into small pieces mechanically is more beneficial than chemical method of treating the cells with enzyme.
3. Feeder cells are not essential for maintaining the hPSCs. They could be replaced by Matrigel-coated culture dish, though more expensive. But MEF cells contribute to a better long-term maintenance of hPSCs under undifferentiated condition.

2.3.2 Establishment of mFLSCs

mFLSCs are prepared as described, and the procedure has been depicted in detail [26, 27, 34, 40]. We simplify the description of its establishment process here. Briefly, mFL are removed carefully from embryonic day15 (E15) C57/Black6 mice. After triturating with two glass slides carefully, mFL total cells were washed once with D-PBS (-) and plated in a gelatin-coated 10-cm dish at a density of 2–3 fetal livers per dish. After 2 days in culture, floating cells are removed through washing with D-PBS (-) gently, and add fresh mFLSCs culture medium. When reaching a confluence after 4–5 days, the mFLSCs are treated with 0.05 % trypsin/EDTA and replated into a new culture dish at a rate of 1:1.5–1:2. After enough quantity has been gained, harvest the cells and cryopreserve them in liquid nitrogen.

Notes:

1. The efficiency of production of cell lines from mFLSCs is low at large, so it should be prepared very carefully.
2. Make sure to tear off the fascia tissue on the fetal livers before triturating.
3. Do not touch the culture dish within the first 2 days. Disturbance before mFLSCs completely adhere to surface of culture dish will damage cells irreversibly.
4. Good quality of P1 mFLSCs is composed of a loosely distributed stromal cell layer. They are not easily to be occupied by narrow and paralleling fibroblastic cells.
5. After P1, the proliferation of the mFLSCs decreases down drastically over the passaging, especially when frozen mFLSCs are thawed and recultured. In this situation, a reduction in area ratio should be considered, and the addition of bFGF (5–10 ng/mL) and increasing the percentage of FBS up to 20–30 % in culture medium are beneficial for the proliferation of mFLSCs.

2.3.3 Maintenance of mAGMS-3 Cell Line

The mAGMS-3 cell line has been established successfully in our laboratory [32] and is maintained in liquid nitrogen. It can strongly induce hematopoietic differentiation of hPSCs by coculture with them.

1. Thaw one vial of cells cryopreserved in liquid nitrogen and plated them on gelatin-coated culture dish. mAGMS-3 cells reach confluence by 2–3 days in culture regularly. Passage these cells by treating with 0.05 % trypsin/EDTA and replating in a new culture dish.
2. After passage 3–5 times, a large quantity of mAGMS-3 cells can be harvested. According to the need of experiments, sufficient cells are cryopreserved in liquid nitrogen as working library, at a density of 1×10^6 cells per cryogenic vial.
3. Before coculture, mAGMS-3 cells are thawed and passaged one time as described above.

Notes:

1. When the mAGMS-3 cells reach confluence in culture, treat them with 0.05 % trypsin/EDTA, and replated into a new culture dish, the appropriate ratio for the passage is about 4–6 in area.
2. When passage, the time of digestion process and the seeding density (20–30 % initial confluence) are important factors to keep the mAGMS-3 cells maintained with primary feature.
3. mAGMS-3 cells are prone to adhere to each other in serum-containing solution. So trypsin solution should be diluted by D-PBS (-) (4 °C) first to avoid

accumulating, and the enzyme should be terminated by serum-containing medium after that.

4. Observed under phase contrast microscope, the naïve cells have clear boundaries and the aging cells become larger and more flat.

2.3.4 Coculture of Undifferentiated hPSCs with mFLSCs/mAGMS-3

1. Before coculture, plate the mFLSCs or mAGMS-3 cells in gelatin-coated 6-well culture plate at $1–2 \times 10^5$ per 6-well plate. After about 1–2 days, the fetal stromal cells reach a good confluence.

2. Radiate the confluent mFLSCs at 25 Gy and mAGMS-3 cells at 15 Gy. After radiation, the stromal cells can be utilized at anytime within 3 days. The culture medium should be exchanged to hPSCs-maintaining medium 1–2 h right before hPSCs are plated.

3. For inducing hematopoietic cells, undifferentiated hPSCs colonies should be mechanically picked up because the mass size of the colony is very much related with the efficiency of blood cell production. For this reason, the colonies are first cutted into small pieces by using a 200-μL pipette tip under a reverse microscope in a clean bench. Individual mass of the undifferentiated hPSC colony was then carefully moved onto the culture plates prepared with radiated mFLSCs/mAGMS-3 cells. In order to obtain high efficiency, routinely each piece of the undifferentiated hESC colony should contain about $0.5–1 \times 10^3$ cells.

4. Plate the undifferentiated hPSCs-colony pieces onto the irradiated stromal cells at about 20–30 pieces per one 6-well. The medium is hPSCs-maintaining medium containing 5 nmol Y-27632 in order to increase the adherent rate of hPSCs pieces.

5. 1 mL hPSCs-maintaining medium is added lightly into per one 6-well plate on the 1st day, and fresh medium is changed on the 2nd day.

6. The medium is exchanged to hematopoiesis-inducing medium on the 3rd day when the hPSCs colonies are growing bigger and still keep undifferentiated. And it is recorded as day 0 of differentiation. Then, the medium should be changed every 1–2 days.

7. During the first 4–6 days after changing with the hematopoiesis-inducing medium, the hPSCs colonies grow larger and begin to differentiate, with the outer edge of the colonies looking like mesoderm structure.

8. Around days 6–8, the cobble stone-like cells will appear, and the proliferation of the cells accelerates which lead to the pH value change rapidly. To moderate the pH change, the medium should be changed a little more frequently or by reducing FBS concentration in the medium. These means could help to slow

down the acidification process of the medium and provide a stable culture microenvironment which is beneficial for the early hematopoietic differentiation.

9. After 10 days in coculture, the hematopoietic CD34$^+$ cells amplify rapidly and reach peak on day 14. Then, the hematopoiesis-supporting potential decreases and exhausts gradually after 16 days. In order to further inducing them toward mature blood cells, the HPCs should be shifted to a new environment for further development and maturation.

Notes:

1. The maintaining status of mAGMS-3 is one of the crucial factors in determining the hematopoietic efficiency of the coculture system.
2. The confluence of the stromal cells should in harmony with the size of hPSCs pieces. About 85–90 % confluence of the stromal cells and about 0.5–1 × 10^3 cells in each hPSCs piece are favorable to an efficient hematopoietic differentiation. According to our experiences, the larger and thicker undifferentiated hPSCs colonies have better hematopoietic efficiency when coculture with the stromal cells than small and thin ones do.
3. Use hPSCs-maintaining medium of the first 3 days to help the hPSCs adapt to the new matrix smoothly.
4. Although VEGF (20 ng/mL) in hematopoiesis-inducing medium is not necessary for differentiation, it can improve the generation of CD34$^+$CD45$^+$ cell fraction [41].
5. Components of FBS are crucial for the differentiating efficiency. Thus, the hematopoietic supporting potential of different commercial FBS should be compared through colony formation analysis by human cord blood (hCB) CD34$^+$ cells or by hPSC-mFLSCs/mAGMS-3 coculture cells. It is found that significant differences of hematopoietic inducing ability exist in even different batches of FBS products.
6. After 10 days of coculture, medium should be added softly and slowly to avoid impairing the microcavity structure where hematopoietic cells originate and proliferate.

2.3.5 Colony Formation Analysis of Hematopoietic Progenitors Derived from hPSCs Coculture with mFLSCs/mAGMS-3

The technique of colony assay was established by Ogawa and Nakahata [42–44]. The potential of hematopoiesis can be evaluated subtly by this method. The individual hematopoietic colony is derived from single cell and specifies a lineage potential at a clonal level. Most of the myeloid cells and erythrocytes can be stimulated in the semisolid colony culture system by adding a cytokine cocktail to favor hematopoietic cell development. Our previous work has showed that considerable hematopoietic progenitors including erythroid progenitors could be generated in this system [27]. Routinely, a 4 or 5 mL mixture of semisolid cell culture should be prepared for one test. The components needed are listed as below:

- α-methylcellulose (final concentration: around 1 %),
- BSA solution (final 1 %),
- FBS (heat inactivated, final 30 %),
- 2-ME (final 10^{-4} M),
- rhSCF (final 100 ng/mL),
- rhIL-3 (final 10 ng/mL),
- rhIL-6 (final 50 ng/mL),
- rhFL (final 10 g/mL),
- rhTPO (final 50 ng/mL),
- rhEPO (final 4 IU/mL), and
- rhG-CSF (final 10 ng/mL).

1. The total coculture-derived cells (or sorted fractionated populations) on different culture days (D8–D18) are harvested and plated to colony cultures, at a density of about 2–5×10^5 total cells or 2–5×10^3 CD34$^+$ cells in 1 mL semisolid culture.
2. Mix the culture well using a 2-mL syringe with an 18-G needle and then shake it violently to mix further.
3. The culture should be put away on ice for 20 min before plating to let the bulbs float up.
4. Lift accurate 1 mL of the culture into one 35-mm Petri dish. The culture is so viscous that only 4 mL aliquots can be made from 5 mL preparation.
5. Two culture dishes and another with sterile water are put into a 10-cm dish together, and then, they are cultured at 37 °C in a 5 % CO_2 high humidity incubator.
6. Observation and calculation of colonies should be performed around days 7–14. The criterion for identifying colony types has long been established in our laboratory [27, 42–44].
7. Pick up large E-burst colonies from the methylcellulose culture, wash them with α-MEM and then replate the erythroid cells in a suspension culture or cytospin them on glass slides directly for next detection.

Notes:

1. Do not move the cultures until culture for 5 days, shaking may leads to several daughter colonies come from one individual colony. It may effect colony counting and give rise to wrong result at last.
2. Erythroid cell colonies include E colonies, E bursts, and Mix colonies, which could be defined refer to reference [27]. The numbers of E colonies are determined on days 7–10 in culture, while the numbers of E bursts and Mix colonies are counted on days 12–14 in semisolid culture.

2.3.6   Erythrocytes Derived from hPSCs Coculture with mFLSCs/mAGMS-3

2.3.6.1   Expansion of Erythroid Progenitors

1. Wash the cocultures by D-PBS (-) twice and treat them by 0.25 % trypsin/EDTA solution.
2. After centrifuge, harvested cells are resuspended in SF-RBC + 7FCs medium and plated in ultra-low adherent culture 6-well plate, including about $1-2x10^6$ total cells per well.
3. The medium should be changed every 2–3 days, and after 5–7 days, the HSCs/HPCs are expanded looking like grape cluster-like structure, while most other cells died. The culture medium is exchanged to SF-RBC + 3FCs medium.

Notes:

1. The efficiency of the production of multipotential HPCs in the cocultures overwhelmingly decides the later erythrocyte production.
2. The hematopoietic structure in cocultures is compact. For better digestion, the cocultures should be immersed in D-PBS (-) for 2–3 min to eliminate the remaining serum in the structure and the 0.25 % trypsin/EDTA solution should be prewarmed to 37 °C before use.
3. According our experiences, total harvested coculture cells should be plated into the following culture system, because cells other than hematopoietic lineage in the cocultures might help to promote a better expansion of hematopoietic progenitors.
4. Although cell number of CD34$^+$ fraction reaches peak on day 14 coculture, the erythroid progenitors emerge much earlier than that. According to our data, the cocultures on day 10–12 have a better erythrocyte production after suspension culture.

2.3.6.2   Differentiation and Maturation of Erythrocytes

1. The cells in suspension culture on day 6 are centrifuged at 1200 rpm for 5 min and then resuspend with SF-RBC medium containing EPO (final: 2 IU/mL) only. Change the medium every 3–4 days if the medium shows yellow color.
2. In liquid culture, more and more erythroid cells are expanded. After 20 days, the purity of erythrocytes reaches up to 95 %.

Notes:

The erythrocytes in this stage are more and more fragile over time. Try best not to centrifuge or at a low centrifugation speed to protect cells when change the medium.

2.3.7 Characterization of hPSC-Derived Erythrocytes by Flow cytometry Assay

Erythroid cells are originated from hPSCs coculture with mFLSCs/mAGMS-3, and then, erythrocytes are expanded and matured in suspension culture. It is known that mature human erythrocytes from hCB-CD34$^+$ cells are GPA$^+$/CD34^{neg}/CD45^{neg}/CD71^{medium}/CD47^{neg}/EPO-R^{neg} [12, 45]. To demonstrate the phenotypic expression pattern of erythrocytes, erythroid cells at different inducing stages are detected by flow cytometry assay. A representative flow cytometry data of erythroid cells derived from hESC (H1)/mAGMS-3 coculture are shown in Fig. 2.3. Based on our data, the erythrocytes generated in our coculture system are typically GPA$^+$CD45^{neg}, and the phenotypic pattern of the mature erythrocytes (by day 12 + 24 H1/mAGMS-3) is similar to that of the adult-type mature erythrocytes.

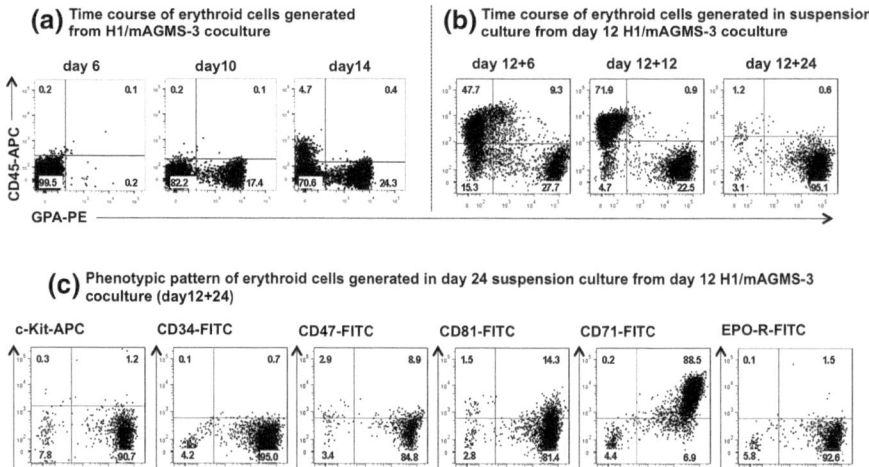

Fig. 2.3 Representative flow cytometry profiles of erythroid cells derived from hESC (H1)/mAGMS-3 coculture. **a** CD45 and GPA co-expression overtime of cells derived from H1/mAGM-S3 coculture. **b** CD45 and GPA co-expression overtime of erythroid cells of suspension culture stage derived from day 12 H1/mAGM-S3 coculture. Abundant mature erythrocytes (GPA$^+$CD45$^-$) are generated. **c** The c-kit, CD34, CD47, CD81, and CD71 expression pattern of erythrocytes derived from day 12 + 24 H1/mAGMS-3 coculture

1. The total harvested coculture cells are incubated with normal rabbit serum. Add 10 µL rabbit serum to 1×10^6 cells.
2. The total harvested coculture cells are distributed into $0.2\text{--}0.5 \times 10^6$ per sample in 0.1 mL volume.
3. Add monoclonal antibodies that conjugated with FITC, PE, or APC. 7-AAD is used to exclude dead cells. Cells are stained on ice in dark for 30 min.
4. Wash the incubated cells twice with SM, and resuspend them with 0.1–0.2 mL SM solution. Then, filter the cell suspension through a 38-µm nylon membrane to remove cell masses.
5. The stained samples are analyzed on FACS Canto™II (Becton Dickinson Company). The cells are gated by SSC and FSC first, and then, the living cells are gated by 7-AAD negative fraction. Record of the data can be further analyzed by BDFACSDiva software or FlowJo software (Version 7.2.5).

Notes:

1. The GPA antibody (Clone, JC159) from DAKO company is more sensitive for detecting embryonic and fetal erythroid cells than others.
2. The CD45 antibody (Clone, 2D1) from DAKO company is more sensitive to embryonic hematopoietic cells than others.

2.3.8   Hb Components of hPSC-Derived Erythrocytes Detected by Immuno-Staining Assay

The various globins are encoded by two gene clusters, the α cluster on chromosome 16 encodes including the embryonic ζ-globin and adult α-globin, and the β cluster on chromosome 11 encodes containing the embryonic ε-globin, the fetal γ-globin, and adult δ- and β-globins [46]. There are two accompanying switches, a switch from embryonic-to-fetal globins early in gestation and then from fetal-to-adult globins around the time of birth. The Hb content is a good criterion to evaluate the maturity of erythrocytes. We detect the Hb components of hPSC-derived erythrocytes by immuno-staining assay as reported [27]. Erythrocytes derived from hPSCs by our coculture system are able to mature progressively and overwhelmingly express high rate of adult-type β-globin (Figs. 2.2d and 2.4d).

1. Circle the cells cytospined on the glass slides using the liquid blocker pen.
2. Fix the cells with 4 % PFA solution at 4 °C for 30 min and then wash the cells with D-PBS (-) for three times.
3. Treat the membrane of cells with 5 % SM/0.1 % Triton-100/D-PBS (-) at 4 °C for 30 min to degrade the lips on the membrane, block the unspecific binding sites on/in cells, and then wash with 5 % SM/D-PBS (-) for three times.
4. Incubate the cells with the primary Abs Hb [1: 100 diluted with 5 % SM/D-PBS (-)] at 4 °C over night.

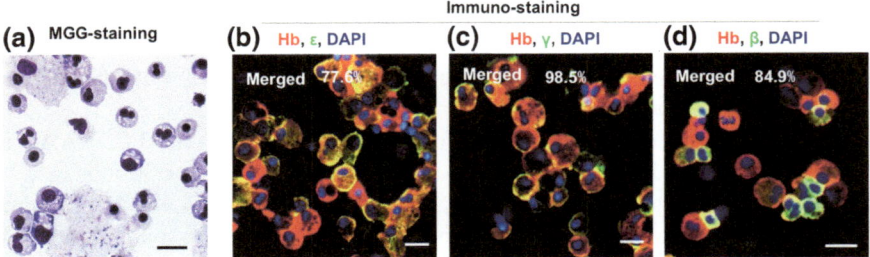

Fig. 2.4 Characteristics of erythroid cells generated from hESC (H1)/mAGMS-3 coculture on day 12 + 24. **a** Photographs of MGG staining. (60× magnification; bar = 20 μm). **b–d** The Hb components of erythrocytes derived from day 12 + 24 H1/mAGMS-3 (*red* Hb; *green* ε-, γ-, β-globin; *blue* DAPI; 60× magnification; bar = 20 μm)

5. After wash with 5 % SM/D-PBS (-) for three times, stain the cells with Cy3-conjugated secondary Abs [1: 100 diluted with 5 % SM/D-PBS (-)] at room temperature (RT) for 30 min and then wash with 5 % SM/D-PBS (-) and 0.1 % Tween-20/D-PBS (-) for 3 times, respectively.
6. Stain the nucleus with DAPI (0.1 μg/mL) at RT for 5 min and then wash with D-PBS (-) for 3 times.
7. Dispense one drop of mounting medium on to the section to mount cells. Cover slips can be permanently sealed around the perimeter with nail polish.
8. Observe with a fluorescence microscope (Olympus, BX53), count the percentage of Hb and ε-, γ-, β-globin positive cells, and then take photographs by using a CellSens XV image processing software (Olympus).

Notes:

1. Optimal working dilutions of primary and secondary antibodies should be determined experimentally.
2. Both of negative and positive controls should be set in every test. They are the standard for judging negative or positive expressing cells.
3. Washing technique after incubation with primary and secondary is very important. Excessive and insufficient wash will lead to false negative and false positive results, respectively.
4. Dispense one drop of mounting medium on each cover slip and then cover the slides onto the cytospined area one by one.
5. Store mounted slides at 4 °C and protect from light.

2.4 Future Perspects

Because of their distinctive properties for erythrocytes in the progressive maturation from embryonic stage to fully mature adult-type RBCs, the Hb-switching pattern has long been served as classical standard to characterize the maturity of erythrocytes derived from embryonic hematopoiesis. Based on the in vitro developmental model of hPSC-derived erythrocytes, clues for human early hematopoiesis are uncovered partly. Recent reports show that expression of GPA (CD253a), previously known as a typical lineage-specific marker for erythrocytes, defines an initiation of a novel earliest population enriched with definitive multipotential hematopoietic progenitors [47–50].

However, there is still a long way to challenge toward clinical translation. First, a more efficient large-scale manufacturing method should be optimized to significantly save the prohibitively expensive cost of culturing cells in vitro. Many efforts have been made for this purpose, such as optimizing the formula of differentiation medium [51], using semipermeable membranes to save complementary molecules in medium [5], replacing cytokines with cheaper mimetics [52], adding specific small molecules to increase expansion folds of erythroid precursors [53], establishing immortalized erythroid progenitor cell lines [54–57], and using genetic manipulation to upregulate erythroid commitment and amplification [55, 58]. Secondly, there are still some obfuscations in the pivotal regulatory axis of erythropoiesis. Especially, it requires practical solution to the puzzles of enucleation [59] and Hb switching [60] in erythrocytes developed from hPSCs. The synthetic three-dimensional (3D) system to mimic hematopoietic niches [61–64] might be a promising means. New-type polymer materials and nanomaterials should also be explored in this field. Thirdly, in order to fulfill successful therapeutic application, an in vivo model of efficient transplantation and functional assay for hPSC-derived erythrocyte progenitors in immuno-deficiency mice should be established. In addition, although the membrane structure molecules of adult erythrocytes has been studied systematically by Mohandas N and Xiuli An group [45, 65, 66], little is known about the characteristics of embryonic and fetal erythrocyte membrane. Both humanized culture system [67] and animal evaluating models [68] must be developed before clinical translation of the hPSC-derived RBCs.

Since the erythrocytes derived from EBs are of primitive properties, mimicking the erythropoiesis in yolk sac, it is essential to establish an efficient culture system to produce large-scale definitive erythrocytes from hPSCs. To our experiences, large quantity of functionally mature erythrocytes can be reproducibly generated from hPSCs by coculture with murine mFLSCs/mAGMS-3 stromal cell lines derived from normal fetal mice [27]. The culture models established in our laboratory provide an ideal platform to further promote the study on uncovering the mechanisms controlling human early erythropoiesis and erythropoiesis-related diseases and the clinical application of hPSC-derived cRBCs.

References

1. WHO. Blood safety and availability. 2014.
2. Migliaccio AR, Whitsett C, Papayannopoulou T, Sadelain M. The potential of stem cells as an in vitro source of red blood cells for transfusion. Cell Stem Cell. 2012;10:115–9.

3. Neildez-Nguyen TM, Wajcman H, Marden MC, Bensidhoum M, Moncollin V, Giarratana MC, Kobari L, Thierry D, Douay L. Human erythroid cells produced ex vivo at large scale differentiate into red blood cells in vivo. Nat Biotechnol. 2002;20:467–72.
4. Douay L, Giarratana MC. In vitro generation of mature and functional human red blood cells: a model with multidisciplinary perspectives. Bulletin de l'Academie nationale de medicine. 2005;189:903–13 (discussion 914-905).
5. Douay L, Andreu G. Ex vivo production of human red blood cells from hematopoietic stem cells: what is the future in transfusion? Transfus Med Rev. 2007;21:91–100.
6. Giarratana MC, Rouard H, Dumont A, Kiger L, Safeukui I, Le Pennec PY, Francois S, Trugnan G, Peyrard T, Marie T, Jolly S, Hebert N, Mazurier C, Mario N, Harmand L, Lapillonne H, Devaux JY, Douay L. Proof of principle for transfusion of in vitro-generated red blood cells. Blood. 2011;118:5071–9.
7. Flores-Guzman P, Fernandez-Sanchez V, Mayani H. Concise review: ex vivo expansion of cord blood-derived hematopoietic stem and progenitor cells: basic principles, experimental approaches, and impact in regenerative medicine. Stem Cells Transl Med. 2013;2:830–8.
8. Jing Q, Cai H, Du Z, Ye Z, Tan WS. Effects of agitation speed on the ex vivo expansion of cord blood hematopoietic stem/progenitor cells in stirred suspension culture. Artif Cells Nanomed Biotechnol. 2013;41:98–102.
9. Migliaccio AR, Whitsett C, Migliaccio G. Erythroid cells in vitro: from developmental biology to blood transfusion products. Curr Opin Hematol. 2009;16:259–68.
10. Thomson JA, Itskovitz-Eldor J, Shapiro SS, Waknitz MA, Swiergiel JJ, Marshall VS, Jones JM. Embryonic stem cell lines derived from human blastocysts. Science. 1998;282:1145–7.
11. Takahashi K, Tanabe K, Ohnuki M, Narita M, Ichisaka T, Tomoda K, Yamanaka S. Induction of pluripotent stem cells from adult human fibroblasts by defined factors. Cell. 2007;131:861–72.
12. Migliaccio G, Di Pietro R, di Giacomo V, Di Baldassarre A, Migliaccio AR, Maccioni L, Galanello R, Papayannopoulou T. In vitro mass production of human erythroid cells from the blood of normal donors and of thalassemic patients. Blood Cells Mol Dis. 2002;28:169–80.
13. Mali P, Ye Z, Hommond HH, Yu X, Lin J, Chen G, Zou J, Cheng L. Improved efficiency and pace of generating induced pluripotent stem cells from human adult and fetal fibroblasts. Stem Cells. 2008;26:1998–2005.
14. Papapetrou EP, Lee G, Malani N, Setty M, Riviere I, Tirunagari LM, Kadota K, Roth SL, Giardina P, Viale A, Leslie C, Bushman FD, Studer L, Sadelain M. Genomic safe harbors permit high beta-globin transgene expression in thalassemia induced pluripotent stem cells. Nat Biotechnol. 2011;29:73–8.
15. Sebastiano V, Maeder ML, Angstman JF, Haddad B, Khayter C, Yeo DT, Goodwin MJ, Hawkins JS, Ramirez CL, Batista LF, Artandi SE, Wernig M, Joung JK. In situ genetic correction of the sickle cell anemia mutation in human induced pluripotent stem cells using engineered zinc finger nucleases. Stem Cells. 2011;29:1717–26.
16. Hanna J, Wernig M, Markoulaki S, Sun CW, Meissner A, Cassady JP, Beard C, Brambrink T, Wu LC, Townes TM, Jaenisch R. Treatment of sickle cell anemia mouse model with iPS cells generated from autologous skin. Science. 2007;318:1920–3.
17. Cheng L. Human stem cell models for red blood disease modeling and treatment. In 2014 International symposium on erythrocyte biology, Zhengzhou, China. 2014.
18. Lu SJ, Feng Q, Park JS, Vida L, Lee BS, Strausbauch M, Wettstein PJ, Honig GR, Lanza R. Biologic properties and enucleation of red blood cells from human embryonic stem cells. Blood. 2008;112:4475–84.
19. Miharada K, Hiroyama T, Sudo K, Nagasawa T, Nakamura Y. Efficient enucleation of erythroblasts differentiated in vitro from hematopoietic stem and progenitor cells. Nat Biotechnol. 2006;24:1255–6.
20. Lapillonne H, Kobari L, Mazurier C, Tropel P, Giarratana MC, Zanella-Cleon I, Kiger L, Wattenhofer-Donze M, Puccio H, Hebert N, Francina A, Andreu G, Viville S, Douay L. Red

blood cell generation from human induced pluripotent stem cells: perspectives for transfusion medicine. Haematologica. 2010;95:1651–9.

21. Chang CJ, Mitra K, Koya M, Velho M, Desprat R, Lenz J, Bouhassira EE. Production of embryonic and fetal-like red blood cells from human induced pluripotent stem cells. PLoS ONE. 2011;6:e25761.

22. Chang KH, Nelson AM, Cao H, Wang L, Nakamoto B, Ware CB, Papayannopoulou T. Definitive-like erythroid cells derived from human embryonic stem cells coexpress high levels of embryonic and fetal globins with little or no adult globin. Blood. 2006;108:1515–23.

23. Kobari L, Yates F, Oudrhiri N, Francina A, Kiger L, Mazurier C, Rouzbeh S, El-Nemer W, Hebert N, Giarratana MC, Francois S, Chapel A, Lapillonne H, Luton D, Bennaceur-Griscelli A, Douay L. Human induced pluripotent stem cells can reach complete terminal maturation: in vivo and in vitro evidence in the erythropoietic differentiation model. Haematologica. 2012;97:1795–803.

24. Kaufman DS, Hanson ET, Lewis RL, Auerbach R, Thomson JA. Hematopoietic colony-forming cells derived from human embryonic stem cells. Proc Natl Acad Sci USA. 2001;98:10716–21.

25. Qiu C, Olivier EN, Velho M, Bouhassira EE. Globin switches in yolk sac-like primitive and fetal-like definitive red blood cells produced from human embryonic stem cells. Blood. 2008;111:2400–8.

26. Ma F, Wang D, Hanada S, Ebihara Y, Kawasaki H, Zaike Y, Heike T, Nakahata T, Tsuji K. Novel method for efficient production of multipotential hematopoietic progenitors from human embryonic stem cells. Int J Hematol. 2007;85:371–9.

27. Ma F, Ebihara Y, Umeda K, Sakai H, Hanada S, Zhang H, Zaike Y, Tsuchida E, Nakahata T, Nakauchi H, Tsuji K. Generation of functional erythrocytes from human embryonic stem cell-derived definitive hematopoiesis. Proc Natl Acad Sci USA. 2008;105:13087–92.

28. Dias J, Gumenyuk M, Kang H, Vodyanik M, Yu J, Thomson JA, Slukvin II. Generation of red blood cells from human induced pluripotent stem cells. Stem Cells Dev. 2011;20:1639–47.

29. Qiu C, Hanson E, Olivier E, Inada M, Kaufman DS, Gupta S, Bouhassira EE. Differentiation of human embryonic stem cells into hematopoietic cells by coculture with human fetal liver cells recapitulates the globin switch that occurs early in development. Exp Hematol. 2005;33:1450–8.

30. Peschle C, Mavilio F, Care A, Migliaccio G, Migliaccio AR, Salvo G, Samoggia P, Petti S, Guerriero R, Marinucci M, et al. Haemoglobin switching in human embryos: asynchrony of zeta—alpha and epsilon—gamma-globin switches in primitive and definite erythropoietic lineage. Nature. 1985;313:235–8.

31. Tavian M, Coulombel L, Luton D, Clemente HS, Dieterlen-Lievre F, Peault B. Aorta-associated CD34⁺ hematopoietic cells in the early human embryo. Blood. 1996;87:67–72.

32. Xu MJ, Tsuji K, Ueda T, Mukouyama YS, Hara T, Yang FC, Ebihara Y, Matsuoka S, Manabe A, Kikuchi A, Ito M, Miyajima A, Nakahata T. Stimulation of mouse and human primitive hematopoiesis by murine embryonic aorta-gonad-mesonephros-derived stromal cell lines. Blood. 1998;92:2032–40.

33. Ma F, Kambe N, Wang D, Shinoda G, Fujino H, Umeda K, Fujisawa A, Ma L, Suemori H, Nakatsuji N, Miyachi Y, Torii R, Tsuji K, Heike T, Nakahata T. Direct development of functionally mature tryptase/chymase double-positive connective tissue-type mast cells from primate embryonic stem cells. Stem Cells. 2008;26:706–14.

34. Ma F, Nashihama YG, Yang W, Yasuhiro E, Tsuji K. Differentiation oh human embryonic and induced pluripotent stem cells into blood cells in coculture with murine stromal cells. In Jin KYAS, editor. Human embryonic and induced pluripotent stem cells: lineage-specific differentiation protocols. Clifton: Humana Press; 2011. pp. 321–335.

35. Keller G, Lacaud G, Robertson S. Development of the hematopoietic system in the mouse. Exp Hematol. 1999;27:777–87.

36. Matsuoka S, Tsuji K, Hisakawa H, Xu M, Ebihara Y, Ishii T, Sugiyama D, Manabe A, Tanaka R, Ikeda Y, Asano S, Nakahata T. Generation of definitive hematopoietic stem cells

from murine early yolk sac and paraaortic splanchnopleures by aorta-gonad-mesonephros region-derived stromal cells. Blood. 2001;98:6–12.

37. Muller AM, Medvinsky A, Strouboulis J, Grosveld F, Dzierzak E. Development of hematopoietic stem cell activity in the mouse embryo. Immunity. 1994;1:291–301.

38. Medvinsky A, Rybtsov S, Taoudi S. Embryonic origin of the adult hematopoietic system: advances and questions. Development. 2011;138:1017–31.

39. Ledran MH, Krassowska A, Armstrong L, Dimmick I, Renstrom J, Lang R, Yung S, Santibanez-Coref M, Dzierzak E, Stojkovic M, Oostendorp RA, Forrester L, Lako M. Efficient hematopoietic differentiation of human embryonic stem cells on stromal cells derived from hematopoietic niches. Cell Stem Cell. 2008;3:85–98.

40. Ma F, Wada M, Yoshino H, Ebihara Y, Ishii T, Manabe A, Tanaka R, Maekawa T, Ito M, Mugishima H, Asano S, Nakahata T, Tsuji K. Development of human lymphohematopoietic stem and progenitor cells defined by expression of CD34 and CD81. Blood. 2001;97:3755–62.

41. Takayama N, Nishikii H, Usui J, Tsukui H, Sawaguchi A, Hiroyama T, Eto K, Nakauchi H. Generation of functional platelets from human embryonic stem cells in vitro via ES-sacs, VEGF-promoted structures that concentrate hematopoietic progenitors. Blood. 2008;111:5298–306.

42. Nakahata T, Spicer SS, Cantey JR, Ogawa M. Clonal assay of mouse mast cell colonies in methylcellulose culture. Blood. 1982;60:352–61.

43. Nakahata T, Ogawa M. Identification in culture of a class of hemopoietic colony-forming units with extensive capability to self-renew and generate multipotential hemopoietic colonies. Proc Natl Acad Sci USA. 1982;79:3843–7.

44. Nakahata T, Ogawa M. Hemopoietic colony-forming cells in umbilical cord blood with extensive capability to generate mono and multipotential hemopoietic progenitors. J Clin Invest. 1982;70:1324–8.

45. Liu J, Zhang J, Ginzburg Y, Li H, Xue F, De Franceschi L, Chasis JA, Mohandas N, An X. Quantitative analysis of murine terminal erythroid differentiation in vivo: novel method to study normal and disordered erythropoiesis. Blood. 2013;121:e43–9.

46. Stamatoyannopoulos G. Control of globin gene expression during development and erythroid differentiation. Exp Hematol. 2005;33:259–71.

47. Slukvin II. Deciphering the hierarchy of angiohematopoietic progenitors from human pluripotent stem cells. Cell Cycle. 2013;12:720–7.

48. Choi KD, Vodyanik MA, Togarrati PP, Suknuntha K, Kumar A, Samarjeet F, Probasco MD, Tian S, Stewart R, Thomson JA, Slukvin II. Identification of the hemogenic endothelial progenitor and its direct precursor in human pluripotent stem cell differentiation cultures. Cell Rep. 2012;2:553–67.

49. Sturgeon CM, Ditadi A, Awong G, Kennedy M, Keller G. Wnt signaling controls the specification of definitive and primitive hematopoiesis from human pluripotent stem cells. Nat Biotechnol. 2014;32:554–61.

50. Yoder MC. Inducing definitive hematopoiesis in a dish. Nat Biotechnol. 2014;32:539–41.

51. Olivier E, Qiu C, Bouhassira EE. Novel, high-yield red blood cell production methods from CD34-positive cells derived from human embryonic stem, yolk sac, fetal liver, cord blood, and peripheral blood. Stem Cells Transl Med. 2012;1:604–14.

52. Perugini M, Varelias A, Sadlon T, D'Andrea RJ. Hematopoietic growth factor mimetics: from concept to clinic. Cytokine Growth Factor Rev. 2009;20:87–94.

53. Smith BW, Rozelle SS, Leung A, Ubellacker J, Parks A, Nah SK, French D, Gadue P, Monti S, Chui DH, Steinberg MH, Frelinger AL, Michelson AD, Theberge R, McComb ME, Costello CE, Kotton DN, Mostoslavsky G, Sherr DH, Murphy GJ. The aryl hydrocarbon receptor directs hematopoietic progenitor cell expansion and differentiation. Blood. 2013;122:376–85.

54. Nakamura Y, Hiroyama T, Miharada K, Kurita R. Red blood cell production from immortalized progenitor cell line. Int J Hematol. 2011;93:5–9.

55. Kurita R, Suda N, Sudo K, Miharada K, Hiroyama T, Miyoshi H, Tani K, Nakamura Y. Establishment of immortalized human erythroid progenitor cell lines able to produce enucleated red blood cells. PLoS ONE. 2013;8:e59890.
56. Hirose S, Takayama N, Nakamura S, Nagasawa K, Ochi K, Hirata S, Yamazaki S, Yamaguchi T, Otsu M, Sano S, Takahashi N, Sawaguchi A, Ito M, Kato T, Nakauchi H, Eto K. Immortalization of erythroblasts by c-MYC and BCL-XL enables large-scale erythrocyte production from human pluripotent stem cells. Stem Cell Rep. 2013;1:499–508.
57. Hiroyama T, Miharada K, Kurita R, Nakamura Y. Plasticity of cells and ex vivo production of red blood cells. Stem Cells Int. 2011;2011:195780.
58. Hattangadi SM, Wong P, Zhang L, Flygare J, Lodish HF. From stem cell to red cell: regulation of erythropoiesis at multiple levels by multiple proteins, RNAs, and chromatin modifications. Blood. 2011;118:6258–68.
59. Fujimi A, Matsunaga T, Kobune M, Kawano Y, Nagaya T, Tanaka I, Iyama S, Hayashi T, Sato T, Miyanishi K, Sagawa T, Sato Y, Takimoto R, Takayama T, Kato J, Gasa S, Sakai H, Tsuchida E, Ikebuchi K, Hamada H, Niitsu Y. Ex vivo large-scale generation of human red blood cells from cord blood CD34+ cells by co-culturing with macrophages. Int J Hematol. 2008;87:339–50.
60. Xu J, Shao Z, Glass K, Bauer DE, Pinello L, Van Handel B, Hou S, Stamatoyannopoulos JA, Mikkola HK, Yuan GC, Orkin SH. Combinatorial assembly of developmental stage-specific enhancers controls gene expression programs during human erythropoiesis. Dev Cell. 2012;23:796–811.
61. Guvendiren M, Burdick JA. Engineering synthetic hydrogel microenvironments to instruct stem cells. Curr Opin Biotechnol. 2013;24:841–6.
62. Yuan Y, Tse KT, Sin FW, Xue B, Fan HH, Xie Y, Xie Y. Ex vivo amplification of human hematopoietic stem and progenitor cells in an alginate three-dimensional culture system. Int J Lab Hematol. 2011;33:516–25.
63. Zhu J, Marchant RE. Design properties of hydrogel tissue-engineering scaffolds. Expert Rev Med Devices. 2011;8:607–26.
64. Jiang J, Papoutsakis ET. Stem-cell niche based comparative analysis of chemical and nano-mechanical material properties impacting ex vivo expansion and differentiation of hematopoietic and mesenchymal stem cells. Adv Healthc Mater. 2013;2:25–42.
65. Mohandas N, An X. Malaria and human red blood cells. Med Microbiol Immunol. 2012;201:593–8.
66. Mel HC, Prenant M, Mohandas N. Reticulocyte motility and form: studies on maturation and classification. Blood. 1977;49:1001–9.
67. Migliaccio G, Sanchez M, Masiello F, Tirelli V, Varricchio L, Whitsett C, Migliaccio AR. Humanized culture medium for clinical expansion of human erythroblasts. Cell Transplant. 2010;19:453–69.
68. Hu Z, Van Rooijen N, Yang YG. Macrophages prevent human red blood cell reconstitution in immunodeficient mice. Blood. 2011;118:5938–46.

Chapter 3
Derivation of Megakaryocytes and Platelets from Human Pluripotent Stem Cells

Yanfeng Li, Ying Wang, Linzhao Cheng and Zack Z. Wang

Abstract Megakaryocytes (MKs), rare hematopoietic cells in adult bone marrow, produce platelets that are critical to vascular hemostasis and wound healing. Ex vivo generation of MKs from human induced pluripotent stem cells (hiPSCs) provides a renewable cell source of platelets for treating thrombocytopenic patients and allows a better understanding of MK/platelet biology. The key requirements in this approach include developing a robust and consistent method for the production of functional progeny cells, such as MKs from hiPSCs, and minimizing risk and variation due to the animal-derived products in cell cultures. Here, we describe an efficient system to generate MKs from hiPSCs under a feeder-free and xeno-free condition, in which all the animal-derived products were eliminated. Several crucial reagents were evaluated and replaced with FDA-approved pharmacological reagents, including romiplostim (Nplate®, a thrombopoietin analog), Oprelvekin (recombinant IL-11), and Plasbumin (human albumin). This basic and defined differentiation system provides a platform for our future effort in investigation of regulatory factors and protocol optimization toward generating large numbers of platelets ex vivo.

Keywords Human induced pluripotent stem cells (hiPSCs) · Hematopoietic progenitor cells (HPCs) · Megakaryocytes (MKs) · Platelet

3.1 Introduction

The life span of enucleated platelets in the blood stream is about 10 days; therefore, continual production of platelets from mature megakaryocytes (MKs) is required for normal hemostasis in a healthy human adult [1, 2]. Platelet transfusion is an

Y. Li · Y. Wang · L. Cheng · Z.Z. Wang (✉)
Division of Hematology, Department of Medicine, Johns Hopkins University School of Medicine, 720 Rutland Ave, Ross 1029, Baltimore, MD 21205, USA
e-mail: zwang51@jhmi.edu

© The Author(s) 2015
T. Cheng (ed.), *Hematopoietic Differentiation of Human Pluripotent Stem Cells*, SpringerBriefs in Stem Cells, DOI 10.1007/978-94-017-7312-6_3

effective approach of preventing or treating bleeding and is essential for cancer treatment by radiation and chemotherapy and for treatments involving hemato-poietic stem cell (HSC) and bone marrow (BM) transplantation [3]. During embryonic development, immature MKs are produced in the yolk sac [4]. The adult MKs are derived from HSCs in the BM, where MKs are the largest (50–100 μm) and one of the rarest (~ 0.01 %) cells. A unique characteristic of MKs is endo-mitosis, a unique process of DNA/chromosome replication without cellular mitosis, which leads to the formation of large cells with polyploid DNA (ranging from 4 N to 128 N, compared to 2 N–4 N in diploid cells). Platelets are then generated from the cytoplasm of MKs by the demarcation membrane system, in which platelet-specific granules and organelle contents in cytoplasm are packaged and remodeled as proplatelets and then preplatelets before releasing into the circulation as platelets [5]. Enucleated platelets retain surface molecules on the cell membrane such as HLA and MK-specific markers including CD41/CD61 and CD42a/CD42b complexes. Thrombopoietin (TPO) that activates its receptor c-Mpl to stimulate downstream signaling pathways [6–9] is the primary hematopoietic factor to induce HSC and hematopoietic progenitor cell (HPC) differentiation to MKs and platelets [10]. In addition to its role in MK generation and maturation, TPO-Mpl signaling regulates HSC and HPC expansion [11]. There are other regulators besides TPO-Mpl signaling, however, participating in platelet generation because mice lacking either c-Mpl or TPO still have potential to produce platelets at a lower level [10]. Some cytokines, including IL-3, IL-6, IL-11, SDF-1, and FGF-4, have been identified to be involved in MK generation [12, 13].

Massive production of MKs and platelets ex vivo would be extremely valuable for treating thrombocytopenia, but there are currently two challenges. First, MKs, which are present in the BM but absent in peripheral blood (PB), have to be differentiated from HSCs or HPCs. The current inability to expand HSCs ex vivo, especially in a clinically compliant setting, has hindered the capacity to massively expand MKs, which have limited proliferation potential. Second, platelet produc-tion ex vivo from MKs remains inefficient, and lack of large numbers of MKs as starting materials has made this approach more difficult. In the past decade, human pluripotent stem cells, including embryonic stem cells (hESCs) and human induced pluripotent stem cells (hiPSCs), that are capable of self-renewal and differentiation into all somatic cell types became an attractive cell source. Generation of MKs and platelets from human pluripotent stem cells [14–19], especially from HLA-matched hiPSCs, should provide a renewable cell source to treat thrombocytopenic patients. Although with tremendous promise, platelet generation in vitro from hESCs or other reprogrammed cell sources is inefficient currently, even in the presence of stromal cells such as 10T1/2 and OP9 mouse cell lines, which are commonly used in early studies [15, 16, 19]. Whereas mature MKs have limited expansion capa-bilities, the lineage-committed MK precursors, megakaryoblasts, are proliferative. Generation of robust expandable megakaryoblasts would provide an important cell source of platelets for research and clinical application. A recent study demon-strated that coordinative ectopic expression of *c-MYC*, *BMI1*, and *BCL-XL* genes in hiPSCs produces expandable immature MKs in a procedure including coculture with the mouse 10T1/2 stromal cells. Turning off the overexpression of these genes in the immature MKs results in the production of platelets [20]. However, it would be highly desirable if we could generate large numbers of MKs from hiPSCs in the absence of mouse stromal cells and without the need of manipulation of oncogene expression.

A recent study demonstrated that generation of MKs from HPCs is achievable under feeder-free condition [18]. However, this study only examined hESCs and used animal-derived products such as bovine serum albumin (BSA) [18, 21]. These xenogeneic and undefined reagents often cause low reproducibility and conflict with the strict requirements of clinical or preclinical applications [22, 23]. To search for a robust and efficient culture condition to generate expandable MK progenitors from hiPSCs, we developed a serum-free and feeder-free system of hiPSC differentiation to MKs with a high level of reproducibility. In this two-step differentiation system [21, 24–26] which first generates $CD45^+CD34^+$ definitive HPCs followed by MK differentiation, we were able to effectively generate a cell population enriched for $CD41^+CD42a^+$ megakaryoblasts. Moreover, we also used FDA-approved pharmacological agents to replace TPO and BSA in culture medium, which is important for future clinical applications.

3.2 Materials

3.2.1 Cell Lines

1. Human iPSC line BC1, initially derived from human adult BM hematopoietic cells on mouse embryonic fibroblast (MEF) feeders, was adapted to feeder-free conditions using the E8 medium (Life Technologies, A14666SA) [25, 27].
2. Other human iPSC lines were reprogrammed from PB mononuclear cells (MNCs) using improved episomal vectors, containing Oct4, Sox2, Klf4, c-Myc, Bcl-xL, and SV40 large T antigen, under feeder-free and xeno-free culture conditions [28].

3.2.2 Hematopoietic Progenitor Cells Generation from Human iPS Cells

1. Human iPSCs differentiate medium (SFM): 48 % Iscove's modified Dulbecco medium (IMDM, Gibco, cat No. 21056-023), 48 % Ham's F-12 (Corning Cellgro, 10-080 CV), 0.5 % Plasbumin (human albumin, Amgen), 1 % chemically defined lipid concentrate (Gibco cat No. 35050-061), 2 mM GlutaMAX (Gibco, cat No. 51500-061), 50 μg/ml L-ascorbic acid phosphate magnesium salt (Sigma, cat No. A8960), 437 μM 1-thioglycerol (Sigma, cat No. M-6745),

1 % insulin–transferrin–selenium-X (Gibco, cat No. 11905-031), 10 ng/ml recombinant human (rh) FGF2 (PeproTech, cat No. 100-18B), 10 ng/ml BMP4 (PeproTech, cat No. 120-05), 50 ng/ml SCF (PeproTech, cat No. 300-07), 10 ng/ml VEGF-A (PeproTech, cat No 100-20.), and 20 ng/ml romiplostim (Nplate, Amgen).

2. 96-well round bottom plates (Costar, cat No 3788).
3. Accutase solution (Sigma, cat No. A6964).
4. Dulbecco's phosphate buffer saline (Corning Cellgro, cat No. 21-031-CV).
5. Vitronectin (Invitrogen, A14701SA).
6. EDTA dissociation buffer: Add 500 μl 0.5 M EDTA (Gibco, Cat. #15575-038) and 4.5 g NaCl into 500 ml Ca-free Mg-free PBS (Invitrogen, Cat. #14190). The final concentrations are 0.5 mM EDTA plus 0.9 % NaCl. Sterilize by filtration and store at 4 °C.
7. ROCK inhibitor (Stemgent, Y27632).

3.2.3 Megakaryocytes and Platelets Generation from the HPS

1. 100-μm cell strainers (Falcon, cat No. 352360).
2. 6-well plate (Costar, cat No. 3506).
3. IL-11 (PeproTech).
4. Cell counting chamber slides (Invitrogen, cat No. C1008).
5. Antibody: CD34-APC (BD, cat No. 340667), CD45-Alexa Fluor 700 (BD, cat No. 560566), CD41-APC (BD Biosciences, cat No. 559777), CD42-efluor 450 (eBioscience cat No. 48-0428-42), CD42b-FITC (eBioscience, cat No. 11-0429-41), and CD61-FITC (eBioscience, cat No. 11-0619-42).
6. ADP (Sigma, cat No. A2754).
7. CD62P-PE (P-selectin, BD Biosciences, cat No. 555524).
8. Calcein AM (Invitrogen, cat No. C34858).
9. FACS tubes (BD, cat No. 352054).
10. 0.1 % Paraformaldehyde (PFA).

3.3 Methods

3.3.1 Maintenance and Expansion of hiPSC Lines

Human iPSC line BC1, initially derived from human adult BM hematopoietic cells on MEF feeders, was adapted to feeder-free conditions using the E8 medium (the Essential 8 medium commercialized by Life Technologies) [25, 27]. Cells were maintained in an undifferentiated state and routinely passaged as small clumps using the EDTA method or as single cells after enzymatic digestion by accutase (Sigma-Aldrich). **To enhance single cell survival, 10 μm ROCK inhibitor Y27632 was added in the medium for the first 24 h after seeding**.

3.3.1.1 Preparation of VTN-N-Coated Plates (6-Well Plate)

1. Thaw a VTN-N tube (50 μl) at room temperature.
2. Dilute VTN-N in 7 ml PBS, equally divide on 6-well plate, and store at 4 °C (**it can be used for one week**).
3. Incubate the VTN-N-coated plate at 37 °C for at least 1 h before use.

3.3.1.2 EDTA Passaging (6-Well Plate)

1. Suck out the medium from one well of 6-well plate.
2. Add 1 ml EDTA solution and leave it at room temperature for 2 min.
3. Aspirate EDTA solution and then add 2 ml E8 medium.
4. Pipet up and down for 10 times to dissociate hiPSC colonies by 1-ml tip.
5. Suck out VTN-N solution from the new VTN-N-coated plate.
6. Split hiPSC colonies to new plate by the ratio of 1:4–1:8.
7. Wash the well by 2 ml E8 medium, repeat step 6, and add to final volume 2 ml/well.
8. Add 5–10 μm ROCKi Y27632 and make sure the cell suspension is well mixed.
9. Change the medium daily, by adding 2–3 ml of E8 medium per well until cells get 90 % confluent (usually around 4–5 days) for next passaging.

3.3.2 Generation of MKs from hiPSCs

Human iPSCs were differentiated into definitive CD34+CD45+ HPCs, using the "spin-embryoid body" (spin-EB) method in feeder- and serum-free conditions modified from previously described protocols [24, 26, 29]. A schematic outline of experimental design is shown in Fig. 3.1.

1. Before starting EB, the cells need to be more than 70 % confluence.
2. Suck out medium from one well of 6-well plate.
3. Add 1 ml DPBS solution and then remove it.
4. Add 1 ml accutase for 3 min at 37 °C.
5. Add 3 ml SFM and pipet up and down for 10 times to dissociate hiPSC colonies by 1-ml tip, and **make sure all the cells are single cells**.
6. Spin down the cells using centrifuge at 1000 rpm for 5 min and then resuspend cells in 3–5 ml SFM.

Fig. 3.1 Schematic diagram for hiPSC differentiation into megakaryocytes (MKs) and plateletlike particles (PLPs)

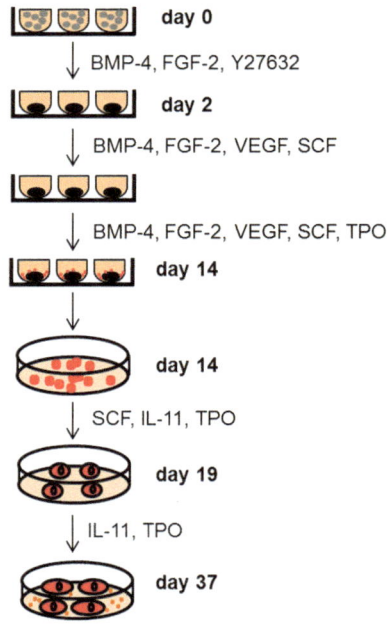

7. Count cell number by cell counting machine.
8. Plate 3000–4000 cells/well into 72 wells (rows B-G) and U-bottom 96-well non-tissue culture plates with 50 μl/well of SFM supplemented with BMP4 (10 ng/ml), bFGF (10 ng/ml), and ROCK inhibitor.
9. Spin the 96-well plates at 1500 rpm for 5 min at room temperature.
10. Fill the other 24 wells on the edges (rows A and F) with 100–200 μl of PBS/sterile water.
11. Place the plates into incubator (This is day 0).
12. On day 2, add 50 μl/well SFM supplemented with BMP4 (10 ng/ml), bFGF (10 ng/ml), VEGF (20 ng/ml), and SCF (100 ng/ml). The final concentration is now BMP4 (10 ng/ml), bFGF (10 ng/ml), VEGF (10 ng/ml), and SCF (50 ng/ml).
13. Add 50 μl/well medium with cytokines (see final concentrations in step 12) every 3 days. Half of the medium can be removed on day 8 before addition of new medium.
14. On day 11, add 50 μl/well SFM supplemented with BMP4 (10 ng/ml), bFGF (10 ng/ml), VEGF (10 ng/ml), SCF (50 ng/ml), and Nplate (20 ng/ml).
15. On day 14, harvest hematopoietic cells present outside EBs by pipetting up and down 5–15 times. Avoid breaking EB down. Pass cells through 100 μm cell strainer (Fig. 3.2a).
16. Spin down the cells with 1000 rpm for 5 min.
17. Resuspend cells in SFM medium supplemented with 3 cytokines. The final concentration is Nplate (50 ng/ml), SCF (20 ng/ml), and IL-11 (20 ng/ml);

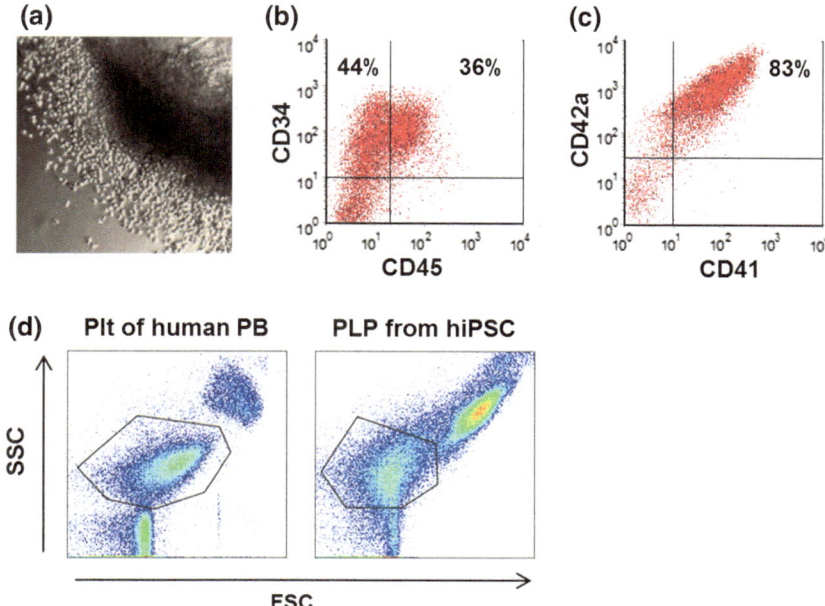

Fig. 3.2 Representative data at different stages of hiPSC differentiation. **a** Image of spin-EB at day 14. **b** Flow cytometry of hematopoietic progenitor cells in suspension on day 14. **c** Flow cytometry of MKs on day 19. **d** Flow cytometry of hiPSC-derived PLPs on day 37 (*left panel*). Platelets of human peripheral blood are used as a control for flow gating (*right panel*)

count cell numbers and put 1 million HPC in 3 ml per well in a 6-well plate. Scale up or down as necessary depending on cell number. Leave some cells for FACS (CD34-PE and CD45-AF700, CD42a-eFluo450 and CD41a-APC) (Fig. 3.2b).

18. Change half or full media every two/three days depending on cell health. Add additional medium if necessary to keep cell density below 5×10^5/ml.

19. Analyze MKs at day 19. FACS analysis of day-19 cells (CD42a-eFluo450, CD41a-APC, CD45-AF700, and CD34-PE; you can detect other MK markers such as CD61 and CD42b at this time). The majority of the cells are MKs (Fig. 3.2c).

3.3.3 Plateletlike Particles Generation from the MKs and Characterization

To generate platelets from MKs, we continued MK culture for additional 18 days in SFM containing IL-11 and Nplate.

1. Resuspend day-19 MKs with SFM containing 10 ng/ml IL and 50 ng/ml Nplate and count cell numbers.
2. Put 1 million MKs into one well of 6-well plate.
3. Change medium twice a week.
4. Harvest the cells on day 37.
5. Spin down them with 100 rpm for 5 min.
6. Collect suspension cells and put them into a new tube.
7. Spin down with 1800 rpm for 10 min.
8. Remove the suspension cells.
9. Resuspend them for FACS to detect plateletlike particles' markers expression (CD42a, CD41) and activation using human PB platelets as a template for gating (Fig. 3.2d). **At this step, we also labeled with calcein AM for live cell staining**.

3.3.4   Functional Analysis of Plateletlike Particles

The activation of hiPSC-PLP was determined by ADP stimulation, and then, CD62P-PE (P-selectin) expression was analyzed by flow cytometry [30, 31].

1. Harvested PLP was resuspended in the buffer with 20 μm ADP stimulation at RT for 10 min.
2. Wash them with 1 ml buffer.
3. Spin down the cells with 1800 rpm for 10 min.
4. Resuspend them and label CD62P and CD41 or CD42a at 4 °C for 30 min.
5. Wash them with FACS buffer.
6. Spin down the cells with 1800 rpm for 10 min.
7. Resuspend them with fix buffer (0.1 % PFA).
8. Run the samples with FACS machine.

3.4   Notes

1. We found that 3000–4000 cells in each 96-well are sufficient to form an EB for most hiPSC lines without polyvinylalcohol (PVA), which has been used to enhance the aggregation of hESCs and formation of EBs by centrifugation [24].
2. The hiPSC-MKs are capable of generating platelets, though the numbers of platelets are still low. The low efficiency of the generation of hiPSC-platelets is likely due to culture system without feeder cells, as in previous studies of human ESCs and iPSCs [15, 20] and primary CD34[+] cells [32].
3. A second generation of TPO receptor agonists, romiplostim (Nplate), has been developed and approved by FDA for clinical applications [33]. The major

reason to choose Nplate over TPO is for a clinical applicable strategy and cost-effectiveness, because Nplate obtained from the clinical research pharmacy is cheaper than recombinant TPO. Nplate is sufficient to substitute for the same dose of TPO and promote hiPSC differentiation to HPCs and MKs.

4. We replaced unreliable sources of BSA by FDA-approved Plasbumin (human albumin), which gives much more consistent results in generation of CD41$^+$CD42a$^+$ MKs.

Acknowledgments This work was partially supported in part by grants from NIH (U01 HL107446 and 2R01 HL-073781) and Maryland State Stem Research Cell Fund (2012-MSCRFII-0124).

References

1. George JN. Platelets. Lancet. 2000;355(9214):1531–9.
2. Kaushansky K. The molecular mechanisms that control thrombopoiesis. J Clin Invest. 2005;115(12):3339–47.
3. Stroncek DF, Rebulla P. Platelet transfusions. Lancet. 2007;370(9585):427–38.
4. Tober J, et al. The megakaryocyte lineage originates from hemangioblast precursors and is an integral component both of primitive and of definitive hematopoiesis. Blood. 2007;109 (4):1433–41.
5. Shultz LD, et al. Human lymphoid and myeloid cell development in NOD/LtSz-scid IL2R gamma null mice engrafted with mobilized human hemopoietic stem cells. J Immunol. 2005;174(10):6477–89.
6. Ramirez PA, Wagner JE, Brunstein CG. Going straight to the point: intra-BM injection of hematopoietic progenitors. Bone Marrow Transplant. 2010;45(7):1127–33.
7. Choi ES, et al. Platelets generated in vitro from proplatelet-displaying human megakaryocytes are functional. Blood. 1995;85(2):402–13.
8. de Sauvage FJ, et al. Stimulation of megakaryocytopoiesis and thrombopoiesis by the c-Mpl ligand. Nature. 1994;369(6481):533–8.
9. Drachman JG, Griffin JD, Kaushansky K. The c-Mpl ligand (thrombopoietin) stimulates tyrosine phosphorylation of Jak2, Shc, and c-Mpl. J Biol Chem. 1995;270(10):4979–82.
10. Bunting S, et al. Normal platelets and megakaryocytes are produced in vivo in the absence of thrombopoietin. Blood. 1997;90(9):3423–9.
11. Fox N, et al. Thrombopoietin expands hematopoietic stem cells after transplantation. J Clin Invest. 2002;110(3):389–94.
12. Avecilla ST, et al. Chemokine-mediated interaction of hematopoietic progenitors with the bone marrow vascular niche is required for thrombopoiesis. Nat Med. 2004;10(1):64–71.
13. Tian X, et al. Bioluminescent imaging demonstrates that transplanted human embryonic stem cell-derived CD34(+) cells preferentially develop into endothelial cells. Stem Cells. 2009;27 (11):2675–85.
14. Takayama N, Eto K. In vitro generation of megakaryocytes and platelets from human embryonic stem cells and induced pluripotent stem cells. Methods Mol Biol. 2012;788:205–17.
15. Lu SJ, et al. Platelets generated from human embryonic stem cells are functional in vitro and in the microcirculation of living mice. Cell Res. 2011;21(3):530–45.
16. Takayama N, et al. Transient activation of c-MYC expression is critical for efficient platelet generation from human induced pluripotent stem cells. J Exp Med. 2010;207(13):2817–30.

17. Gaur M, et al. Megakaryocytes derived from human embryonic stem cells: a genetically tractable system to study megakaryocytopoiesis and integrin function. J Thromb Haemost. 2006;4(2):436–42.
18. Pick M, et al. Generation of megakaryocytic progenitors from human embryonic stem cells in a feeder- and serum-free medium. PLoS ONE. 2013;8(2):e55530.
19. Ono Y, et al. Induction of functional platelets from mouse and human fibroblasts by p45NF-E2/Maf. Blood. 2012;120(18):3812–21.
20. Nakamura S, et al. Expandable megakaryocyte cell lines enable clinically applicable generation of platelets from human induced pluripotent stem cells. Cell Stem Cell. 2014;14 (4):535–48.
21. Ng ES, et al. Forced aggregation of defined numbers of human embryonic stem cells into embryoid bodies fosters robust, reproducible hematopoietic differentiation. Blood. 2005;106 (5):1601–3.
22. Chen G, et al. Chemically defined conditions for human iPSC derivation and culture. Nat Methods. 2011;8(5):424–9.
23. Hulse WL, Gray J, Forbes RT. Evaluating the inter and intra batch variability of protein aggregation behaviour using Taylor dispersion analysis and dynamic light scattering. Int J Pharm. 2013;453(2):351–7.
24. Yahata T, et al. Functional human T lymphocyte development from cord blood CD34$^+$ cells in nonobese diabetic/Shi-scid, IL-2 receptor gamma null mice. J Immunol. 2002;169(1):204–9.
25. Hiramatsu H, et al. Complete reconstitution of human lymphocytes from cord blood CD34$^+$ cells using the NOD/SCID/gammacnull mice model. Blood. 2003;102(3):873–80.
26. Ye Z, et al. Human-induced pluripotent stem cells from blood cells of healthy donors and patients with acquired blood disorders. Blood. 2009;114(27):5473–80.
27. Chou BK, et al. Efficient human iPS cell derivation by a non-integrating plasmid from blood cells with unique epigenetic and gene expression signatures. Cell Res. 2011;21(3):518–29.
28. Chou BK, et al. A facile method to establish human induced pluripotent stem cells from adult blood cells under feeder-free and xeno-free culture conditions: a clinically compliant approach. Stem Cells Transl Med. 2015.
29. Civin CI, et al. Sustained, retransplantable, multilineage engraftment of highly purified adult human bone marrow stem cells in vivo. Blood. 1996;88(11):4102–9.
30. Michelson AD. Flow cytometry: a clinical test of platelet function. Blood. 1996;87(12):4925–36.
31. Michelson AD, Furman MI. Laboratory markers of platelet activation and their clinical significance. Curr Opin Hematol. 1999;6(5):342–8.
32. Cheng L, et al. Human mesenchymal stem cells support megakaryocyte and pro-platelet formation from CD34(+) hematopoietic progenitor cells. J Cell Physiol. 2000;184(1):58–69.
33. Himburg HA, et al. Pleiotrophin regulates the retention and self-renewal of hematopoietic stem cells in the bone marrow vascular niche. Cell Rep. 2012;2(4):964–75.

Chapter 4
Derivation of Functionally Mature Eosinophils from Human Pluripotent Stem Cells

Ya Zhou, Xu Pan, Wenyu Yang, Yanzheng Gu, Bin Mao, Mowen Lai, Wencui Sun, Shu Huang, Tatsutoshi Nakahata and Feng Ma

Abstract The in vitro development of functionally mature blood cells from human pluripotent stem cells [hPSCs; including human embryonic stem cells (hESCs) and induced pluripotent stem cells (hiPSCs)] has proven ideal way to gain information during human early embryonic/fetal hematopoiesis, which never can be mimicked in other species. We recently established an efficient method to produce large quantity of pure and functionally mature eosinophils from hPSCs. The method includes majorly three steps: (1) induction of hematopoietic stem/progenitor cells by coculturing hPSCs with mAGM-3 or mFL stromal cells; (2) large expansion of hPSC-derived hematopoietic stem/progenitor cells and inducing differentiation toward myeloid lineage; and (3) directed differentiation of myeloid progenitor cells into functionally mature eosinophils. The eosinophils induced from hPSCs in our system showed similar morphology and surface marker expression with those in peripheral blood through May-Grünwald-Giemsa staining, transmission electron microscopy (TEM) analysis, flow cytometric analysis (FACS), RT-PCR analysis, and immunofluorescent staining. Furthermore, chemotactic migration and degranulation ability have confirmed the maturity and function of these hPSC-derived

Y. Zhou · X. Pan · B. Mao · M. Lai · W. Sun · S. Huang · F. Ma (✉)
Institute of Blood Transfusion, Chinese Academy of Medical Sciences
and Peking Union Medical College, Chengdu, China
e-mail: mafeng@hotmail.co.jp

W. Yang · F. Ma
State Key Lab of Experimental Hematology, Chinese Academy of Medical Sciences
and Peking Union Medical College, Tianjin, China

Y. Gu · F. Ma
Stem Cell Key Laboratory of Jiangsu Province, Suzhou University, Suzhou, China

T. Nakahata
Center for iPS Cell Research and Application (CiRA), Kyoto University, Kyoto, Japan

F. Ma
Research Center for Stem Cell Therapy, Institute of Blood Transfusion,
Chinese Academy of Medical Sciences and Peking Union Medical College
(CAMS&PUMC), 26 Huacai Road, Longtan Industry Park, Chenghua District,
Chengdu 610052, China

© The Author(s) 2015   51
T. Cheng (ed.), *Hematopoietic Differentiation of Human Pluripotent Stem Cells*,
SpringerBriefs in Stem Cells, DOI 10.1007/978-94-017-7312-6_4

eosinophils. The induction of eosinophils from hPSCs provides us with a perfect model to study the germination, development, differentiation, and maturation of human eosinophiles, which has not been well defined yet. It also provides novel approach to develop patient-tailored therapies by iPSCs for severe allergic diseases as well as deficiencies in early innate immunity. In this review, we will describe the details of methodology for generating these functionally mature eosinophils from hPSCs and the related assay for their function and maturation.

Keywords Human pluripotent stem cells (hPSCs) · Eosinophils · Differentiation · Hematopoiesis · mAGM-3

4.1 Introduction

The production of blood cells from human embryonic stem cells (hESCs) [1] provided valuable in vitro models for analysis of human early hematopoietic lineage commitment and differentiation during embryonic and fetal stages, which never can be mimicked in other species. Recently established human induced pluripotent stem cells (hiPSCs) further enable us to investigate subtly into the mechanism controlling the development of certain diseases with genetic base in a patient-tailored way. The functional blood cells induced from these human pluripotent stem cells (hPSCs) include erythrocytes [2, 3], neutrophil granulocytes [4, 5], megakaryocytes [6], platelets [7], eosinophil granulocytes [8], monocytic and dendritic cells [9], natural killer cells [10, 11], mast cells [11], B lymphocytes, and T lymphocytes [12, 13]. It highlighted the new possibility for these hPSC-derived functional blood cells in various clinical applications such as erythrocytes in transfusion and hematopoietic stem cells (HSCs) in HSC transplantation [14, 15].

There are mainly three approaches to induce hematopoietic differentiation from hPSCs, which include embryoid body (EB) method [16], 2D culture systems [17], and coculture with stromal cell lines [12, 18–20]. In EB method, three-dimensional (3D) spheres of undifferentiated hPSCs are formed, partially representing a very early developing model of human embryos. If with some specific direction toward mesoderm-derived endothelial and hematopoietic differentiation, these 3D EBs can give rise to generation of the first wave of blood cells. On the other hand, a 2D culture system is based on addition of various cytokine combinations without the use of feeder cells. In the coculture system, the typical stromal cell lines used to induce hematopoietic differentiation include the following: OP9, a neonatal murine bone marrow stromal cell line derived from M-CSF OK mice [21]; AGM (aorta/gonad/mesonephros)-derived stromal cell lines such as mAGMS-3 and AGMS-1 [22]; mouse fetal liver stromal cells (mFLSCs) [23]; MS-5 cell lines which are derived from murine BM cells [24]; FH-B-hTERT; and human or mouse marrow stromal cells [19, 20, 25]. It was found that these stromal cells secrete various cytokines needed for the generation of hematopoietic cells. Among these cell lines, the main advantage of mAGM-derived cells is clear by its potential to induce hematopoietic stem/progenitor cells from hESCs with high efficiency [20, 22]. These coculture systems possibly mimic the early development of blood cells in vivo [20].

Having been focusing on hPSC-derived functionally mature blood cells for long, our group has established an efficient method to induce large-scale production of multipotential hematopoietic progenitor cells by coculturing hPSCs with murine fetal hematopoietic niche-derived stromal cells, mFLSC and mAGM cells [2, 23, 26, 27]. In this review, we introduce our protocol to induce functionally mature eosinophils from hPSCs. The protocol includes majorly three steps: (1) induction of hematopoietic differentiation by coculturing hPSCs with mAGM-3 stromal cells (or mFLSCs); (2) expansion of hematopoietic stem/progenitor cells in suspension culture by cytokines; and (3) direct differentiation of multipotential hematopoietic progenitors into mature eosinophils. Using our method, large quantity of highly purified eosinophils can be obtained in the three-step culture system. The eosinophils induced from hPSCs showed similar morphology and surface marker expression with those in peripheral blood through May-Grünwald-Giemsa staining, transmission electron microscopy (TEM) analysis, flow cytometric analysis (FACS), RT-PCR analysis, and immunofluorescent staining. Furthermore, chemotactic migration and degranulation function have confirmed the function and maturity of these hPSC-derived eosinophils. Thus, we have successfully developed a method for efficient production of eosinophils from hPSCs, providing us unlimited supply of human mature eosinophils in vitro. The induction of eosinophils from hPSCs provides us with an ideal model to study the germination, development, differentiation, and maturation of human eosinophils, which largely remains unknown. It also provides us ways to build patient-tailored disease model to uncover the mechanisms controlling those eosinophil-related conditions, such as severe allergic responses, and to screen drugs to cure these diseases.

4.2   Materials

4.2.1   hPSC Lines and Culture Medium

1. hESC lines, H1 and K3, were acquired from the WiCell Research Institute (Madison, WI) and Institute for Regenerative Medicine, Kyoto University (Kyoto, Japan). hiPSC line (253B7) was provided by Professor S Yamanaka at CiRA, Kyoto University (Kyoto, Japan).
2. Matrigel-coated culture dishes (Sumilon, Cat. No. MS-0390 G).
3. hPSC maintaining medium:

   DMEM (GIBCO, Cat. No. 12800-017)
   F12 (GIBCO, Cat. No. 21700-075)
   KSR (GIBCO, Cat. No. 10828-028)

2-ME (Sigma, Cat. No. M7522)
L-Glutamine (GIBCO, Cat. No. 25030-081)
Nonessential amino acid solution (NEAA) (GIBCO, Cat. No. 11140-050)
Basic FGF (GIBCO, Cat. No. 13256-029)

4.2.2 Murine AGMS-3 Cell Line and Fetal Liver Stromal Cells (mFLSCs)

1. Murine AGM: The mAGMS-3 stromal cell line was derived from murine AGM region. The hematopoiesis-supporting potential of mAGMS-3 has not decreased since established in 1998 [28].
2. Gelatin-coated culture dishes (Costar, Cat. No. 3335).
3. mAGMS-3 maintaining medium:

   α-Minimum essential medium (α-MEM) (HyClone, Cat. No. SH30265).
   Fetal bovine serum (FBS) (Biowest, Cat. No. S1580-500) 10 % in volume.

4. mFLSCs were derived from E12.5 to 13.5 fetal liver and cell lines made as reported elsewhere previously [2, 27].

4.2.3 Induction of Multipotential Hematopoietic Progenitor Cells

1. Biological X-ray irradiator (Rad Source Technologies, Inc. RS2000).
2. Undifferentiated hESCs.
3. Radiated mAGMS-3 cells (radiation dose: 13.3 Gy for mAGMS-3)
4. Gelatin-coated 6-well culture plates (Costar, Cat. No. 3335)
5. Hematopoiesis-inducing medium in coculture:

   - Iscove's modified Dulbecco's medium (IMDM) (GIBCO, Cat. No. 12440-053)
   - FBS (HyClone, Cat. No. SH30084.03)
   - Nonessential amino acid solution (NEAA) (GIBCO, Cat. No. 11140-050)
   - 2-ME (Sigma, Cat. No. M7522)
   - Glutamine (Wako, Cat. No. 073-05391)
   - Recombinant human vascular endothelial growth factor (rhVEGF) (Wako, Cat. No. 229-01353)
   - Ascorbic acid (AA) (Sigma, Cat. No. 1043003-1G)
   - Transferrin (Tri) (Sigma, Cat. No. T2252)

4.2.4 *Induction of Eosinophils*

1. Induction Medium I:

   - Iscove's modified Dulbecco's medium (IMDM) (GIBCO, Cat. No. 12440-053)
   - FBS (HyClone, Cat. No. SH30084.03) 10 %
   - rhVEGF (Wako, 229-01353) 16.67 ng/ml
   - rhSCF (BBI, Cat. No. RC015) 100 ng/ml
   - rhIL-6 (BBI, Cat. No. RC007) 100 ng/ml
   - rhTPO (KIRIN, Cat. No. NHK0828-SDM) 5 ng/ml
   - rhFlt3 ligand (BBI, Cat. No. RC024) 5 ng/ml
   - rhIL-3 (KIRIN, Cat. No. MYE0317) 10 ng/ml
   - rhGM-CSF (BBI, Cat. No. RC017) 20 ng/ml

The medium was blended and filtrated through a 0.22-μm filter and then stored in a 4 °C refrigerator. When cytokines were added, the medium should be used within 2 weeks.

2. Induction Medium II:

   - Iscove's modified Dulbecco's medium (IMDM) (GIBCO, Cat. No. 12440-053)
   - FBS (HyClone, Cat. No. SH30084.03)
   - rhVEGF (Wako, 229-01353) 16.67 ng/ml
   - rhSCF (BBI, Cat. No. RC015) 100 ng/ml
   - rhIL-3 (KIRIN, Cat. No. MYE0317) 10 ng/ml
   - rhGM-CSF (BBI, Cat. No. RC017) 10 ng/ml

The medium was blended and filtrated through a 0.22-μm filter and then stored in a 4 °C refrigerator. When cytokines were added, the medium should be used within 2 weeks.

3. Induction Medium III:

   - Iscove's modified Dulbecco's medium (IMDM) (GIBCO, Cat. No. 12440-053)
   - FBS (HyClone, Cat. No. SH30084.03)
   - rhVEGF (Wako, 229-01353) 16.67 ng/ml
   - rhIL-3 (KIRIN, Cat. No. MYE0317) 10 ng/ml
   - rhIL-5 (R&D, Cat. No. AH2310101) 10 ng/ml

The medium was blended and filtrated through a 0.22-μm filter and then stored in a 4 °C refrigerator. When cytokines were added, the medium should be used within 2 weeks.

4.2.5 Flow Cytometric Analysis of hPSC-Derived Eosinophils

1. 0.05 % trypsin/EDTA solution (Wako, Cat. No. 202-16931)
2. Sorting medium (SM):

   - Dulbecco's phosphate-buffered saline without Ca2+ and Mg2+ (D-PBS(-); Wako, Cat. No. 045-29795)
   - 5 % FBS (HyClone, Cat. No. SH30084.03)

3. Antibodies:

   - Anti-human CD34-FITC (BD Biosciences, Cat:555421)
   - Anti-human CD43-FITC (BD Biosciences, Cat:555475)
   - Anti-human CD45-FITC (BD Biosciences, Cat:555482)
   - Anti-human CD81-FITC (BD Biosciences, Cat:551108)
   - Anti-human CD88(C5aR)-FITC (BioLegend, Cat:344306)
   - Anti-human CD81-PE (BD Biosciences, Cat:555676)
   - Anti-human CD88(C5aR)-PE (BioLegend, Cat:344304)
   - Anti-human Siglec-8-PE (BioLegend, Cat:347104)
   - Anti-human CD34-APC (BD Biosciences, Cat:555824)
   - Anti-human CD81-APC (BD Biosciences, Cat:551112)
   - Anti-human Siglec-8-APC (BioLegend, Cat:347104)
   - 7-AAD (BD Biosciences, Cat:51-68981E)

4. Flow cytometry system (Becton Dickinson and Company, BD FACSCantoTM- II

4.2.6 Transmission Electron Microscope (TEM) Analysis of hPSC-Derived Eosinophils

1. 0.1M PBS with 2 % paraformaldehyde and 2.5 % glutaraldehyde
2. Cacodylate buffer
3. 0.1M PBS with 1 % osmium tetroxide
4. 50, 70, 80, 90, and 100 % graded ethanol
5. Epoxy resin
6. Uranyl acetate
7. Reynolds' lead citrate

The sections were inspected using a TEM H7000 (Hitachi, Japan).

4.2.7 May-Grönwald-Giemsa Staining

1. May-Grönwald solution (Merck, HX135575)
2. Giemsa solution (Merck, HX1377334)

4.2.8   Immunochemical Staining

1. 4 % paraformaldehyde (PFA) solution (Boster).
2. Skim milk (BD, REF:232100)
3. D-PBS (-) (Sigma, Cat. No. D5773)
4. Triton X-100 (Uni-Chem, Cat. No. 9002-93-1)
5. DAPI (Roche, REF. 0 236 276 001)
6. Upright fluorescence microscope (Olympus, BX53)
7. Primary Abs

   - Mouse anti-human EPO (Santa Cruz Biotechnology, Cat:E2019)
   - Mouse anti-human MBP (Chemicon Europe, Cat:4192705P3F)
   - Mouse anti-human 2D7 (BioLegend, Cat:346202)
   - Mouse anti-human pro-MBP (BioLegend, Cat:346802)

8. Secondary Abs

   - Donkey anti-goat IgG (H + L)-CyTM3 (IR, Cat:107715)
   - Donkey anti-mouse IgG (H + L)-FITC (IR, Cat:104895)
   - DAPI (Roche, Cat:70217321)

4.2.9   RT-PCR

1. RNA subtract kit (Invitrogen, Cat:12183-016)
2. iScript^{TM} cDNA Synthesis Kit (Bio-Rad, Cat:170-8890)

4.3   Methods

4.3.1   Maintenance of hPSC Lines

1. The hESCs (Line H1) and hiPSC (Line 253B7) were maintained on matrigel as described. Differentiated cell colonies were picked out, and medium was changed everyday. Plates of undifferentiated hPSC cultures were incubated at 37 °C in a humidified atmosphere containing 5 % CO_2.
2. Undifferentiated hPSCs were passaged every 5–7 days.

4.3.2 Coculture of Undifferentiated hPSC with AGMS-3 (or mFLSCs)

1. The mAGMS-3 cells were prepared in gelatin-coated 6-well culture plates. 1–2 × 10^5 cells per well were cultured for 2 days to reach a good confluence of 90–100 %.
2. The stromal cells can be used at any time within 5 days after radiation at 13.3 Gy.
3. Cut the undifferentiated hPSC colonies into small pieces which contain 0.5–1 × 10^3 cells with a pipet tip under a reverse microscope. Poke the small colonies free with a pipet tip.
4. Gently collect the small pieces of hPSC colonies from the plate and wash once with fresh hPSC medium.
5. Plate the 50–70 pieces of undifferentiated hPSC colonies onto radiated AGMS-3 cells. When the undifferentiated hESC colonies grow bigger after 2–3 days, exchange the medium to 10 % FBS-IMDM inducing medium.
6. The hematopoietic cells appeared at the outskirt area of an expanding colony at around day 8–10.
7. After culture day 10, these hematopoietic cells proliferate rapidly and begin to differentiate to myeloid progenitor cells. In our coculture system, the production of the hematopoietic cells reaches a peak on days 12–14 and then gradually decreases in number.
8. Typically, total coculture cells were harvested on day 14 and treated with 0.25 % trypsin/EDTA solution. Cells were filtered once through a 70-mm cell strainer to collect the single-cell suspension for further use.

We omitted the similar method by coculturing hPSCs with mFLSCs.

4.3.3 Induction of Eosinophils

1. 1st Step: Expansion of hematopoietic stem/progenitor cells.
   The suspension cells were recultured in Induction Medium I for 7 days, mainly to expand hematopoietic stem/progenitor cells for further differentiation experiments. The culture medium was renewed every 3 days.
2. 2nd Step: Induction of eosinophil differentiation.
   During the next 7 days, the medium was changed into Induction Medium II, aiming at an efficient induction of eosinophil differentiation. The culture medium was also renewed every 3 days.
3. 3rd Step: Induction of the maturation of hPSC-derived eosinophils.
   During the last 7 days, the cells were cultured in Induction Medium III for 7 days. The medium is mainly used to induce the maturation of those hPSC-derived eosinophils.

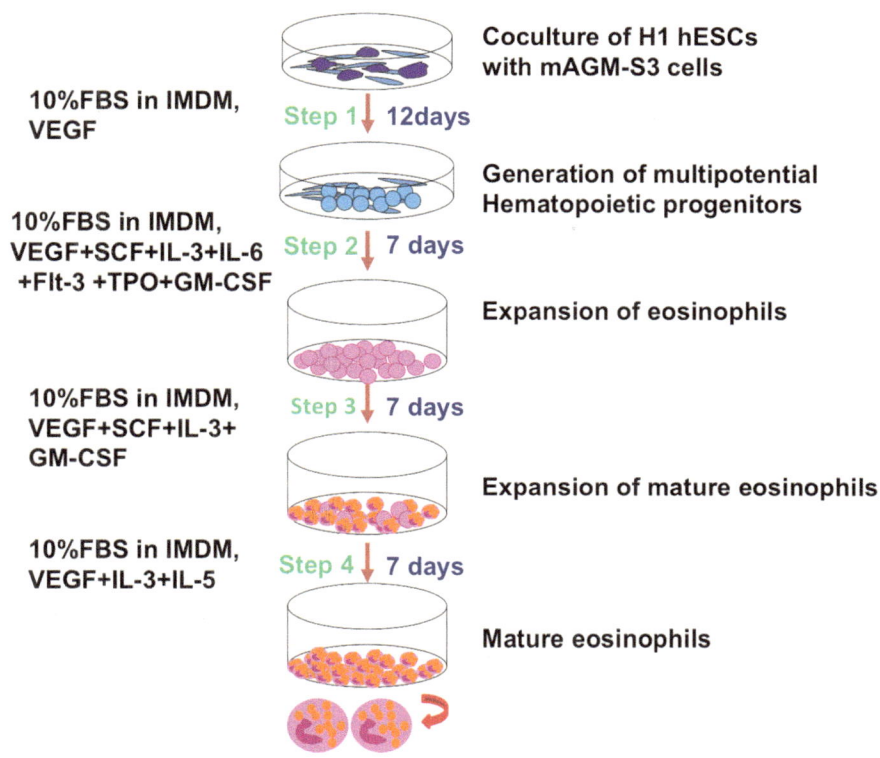

10%FBS in IMDM, VEGF

Step 1 ↓ **12days**

Coculture of H1 hESCs with mAGM-S3 cells

10%FBS in IMDM, VEGF+SCF+IL-3+IL-6 +Flt-3 +TPO+GM-CSF

Step 2 ↓ **7 days**

Generation of multipotential Hematopoietic progenitors

Expansion of eosinophils

10%FBS in IMDM, VEGF+SCF+IL-3+ GM-CSF

Step 3 ↓ **7 days**

Expansion of mature eosinophils

10%FBS in IMDM, VEGF+IL-3+IL-5

Step 4 ↓ **7 days**

Mature eosinophils

Fig. 4.1 Schematic representation of the procedures for maturation of hPSC-derived eosinophils

Figure 4.1 shows a schematic representation of the procedures for generation of mature eosinophils from hPSCs.

4.3.4   Characterization of hPSC-Derived Eosinophils by Flow Cytometric Analysis

1. The suspension cells at day 21 were harvested, and samples (every 1×10^6 cell/vial) were preincubated with 10 μl normal rabbit serum for 30 min to block nonspecific binding on the cell surface.
2. Wash once with SM and resuspend in 0.1 ml SM.
3. The cells were stained for 30 min on ice with various mAbs conjugated with phycoerythrin (PE), allophycocyanin (APC), fluorescein isothiocyanate (FITC), or unconjugated mAbs.
4. Wash once with SM and resuspend in 0.5–1 ml SM per sample. Add 7-AAD to stain the dead cells before analysis.

5. Stained cells were analyzed by a BD FACS Canto-II cytometry system with FlowJo software (version 7.2.5).

A representative FACS data of hPSC-AGMS-3 coculture-derived mature eosinophils are shown in Fig. 4.2a.

4.3.5 Characterization of hPSC-Derived Eosinophils by Transmission Electron Microscope (TEM) Analysis

1. The suspended hPSC-derived eosinophils harvested from day 21 are fixed in 0.1M PBS with 2 % paraformaldehyde and 2.5 % glutaraldehyde for at least 12 h followed by washing with cacodylate buffer for six times.
2. Specimens are then treated with 1 % osmium tetroxide in PBS for 2 h followed by washing with cacodylate buffer for six times.
3. The specimens are then dehydrated using 50, 70, 80, 90, and 100 % graded ethanol (Wako).
4. After dehydration, samples are cleared with propylene oxide and embedded in epon. Thin sections of cured samples are stained with uranyl acetate and Reynolds' lead citrate.
5. The samples are scanned with TEM (Hitachi H7000).

Figure 4.2b shows that the cells under TEM are mainly oval and contain dispersed granules encircled in membrane. However, the hPSC-derived eosinophils harvested at day 21 do not show typical crystal structures.

4.3.6 May-Grönwald-Giemsa Staining

1. The hPSC-derived eosinophils in suspension culture at day 21 are washed with PBS once and centrifuged onto glass slides by cytospin.
2. Centrifuged preparations are fixed with May-Grönwald solution for 4 min followed by washing with running water once for 1 min.
3. Centrifuged preparations are then stained with Giemsa solution for 20 min and then washed by running water once for 2 min.
4. Centrifuged preparations are dried by airing.
5. hPSC-derived eosinophils are typically round or oval in shape under the microscope (Fig. 4.2c). The nucleus is commonly lobulated or rod-shaped and the chromatin agglutinated. The cytoplasm is mauve and contained thick orange granules.

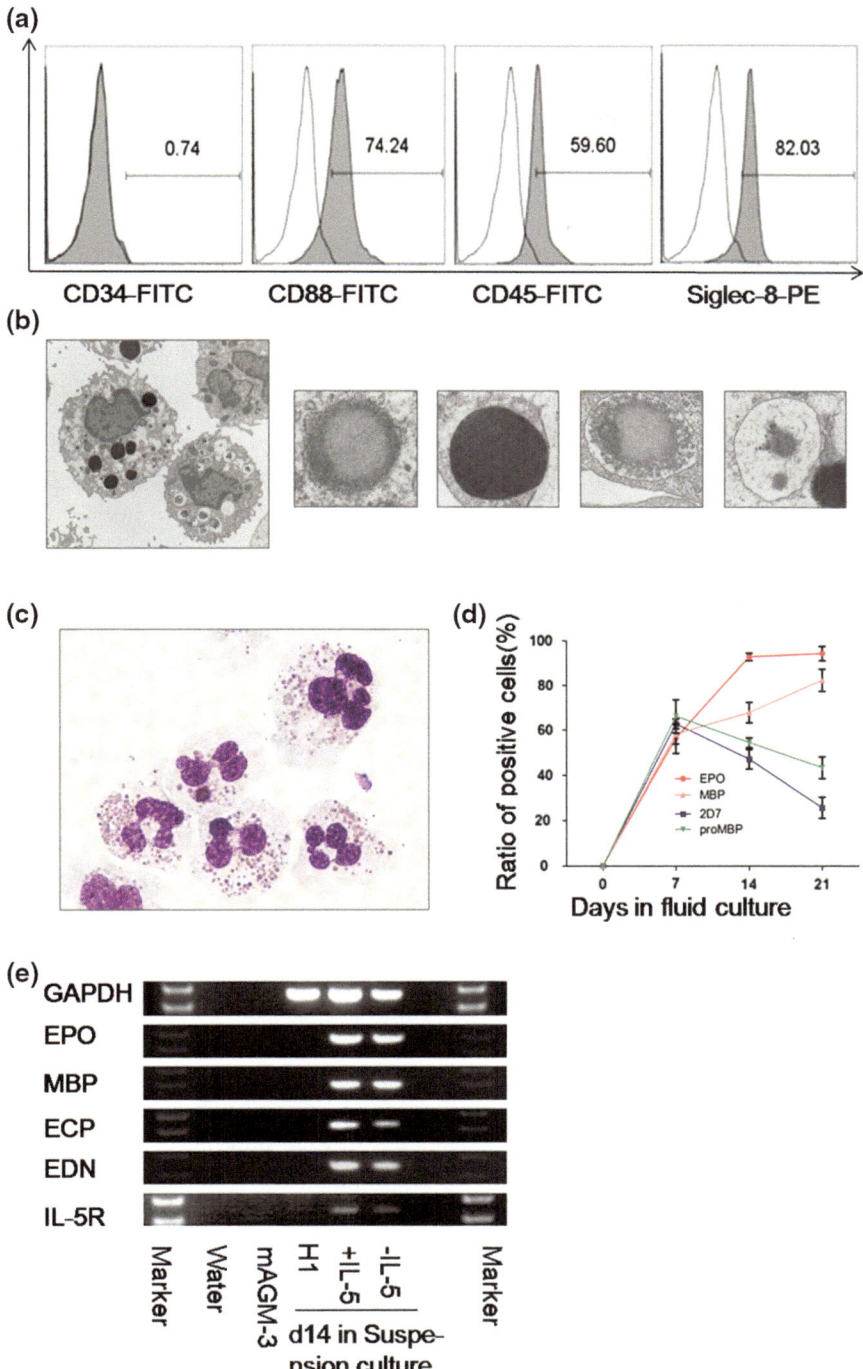

◄ **Fig. 4.2** Eosinophils derived from hESCs in our system **a** FACS analysis of day 28 suspension coculture cells derived from hESCs. **b** Morphology of mature eosinophils by scanning electric microscope. **c** May-Grünwald-Giemsa staining of mature eosinophis (100 ×). **d** The expression of EPO, MBP, 2D7, proMBP1 of cells in fluid culture from day 4 to day28. **e** The expression level of genes specific for mature eosinophils (EPO, MBP, EDN, ECP, IL-5R) by RT-PCR

4.3.7 RT-PCR

1. RT-PCR analysis is applied to detect early stages of hPSC-derived hemato-poiesis and globin expression.
2. Total RNA is prepared from hPSC/mAGMS-3 cocultured cells and hPSC-derived mature eosinophils by RNA subtract kit (Invitrogen, Cat:12183-016).
3. Single-stranded cDNA was synthesized from total RNA using iScript™ cDNA Synthesis Kit (Bio-Rad, Cat:170-8890).
4. PCR conditions are the same as reported. Human gene-specific primers were used throughout our experiments to avoid interference from mAGM cells [8, 16, 19]. Amounts of cDNA template are standardized against the relative expression of GAPDH in each sample.
5. The expression level of genes specific for mature eosinophils (EPO, MBP, EDN, ECP, IL-5R, etc.) is detected by RT-PCR (Fig. 4.2e).

4.3.8 Immunostaining

1. The cells in suspension culture at various times (day 4, 7, 14, 21, 28) were washed with PBS once and centrifuged onto glass slides.
2. Centrifuged preparations were dried by airing.
3. Centrifuged preparations were fixed with cold (4 °C) 4 % paraformaldehyde (PFA) for 30 min followed by washing with PBS for three times.
4. The membrane of the cells was permeabilized with cold (4 °C) PBS containing 5 % skim milk and 0.1 % Triton X-100 for 30 min.
5. The slides were incubated with primary anti-human Abs (mouse anti-human EPO, MBP, 2D7, pro-MBP; dilution at 1:100) in a wet box at 4 °C overnight and then washed with PBS containing 5 % skim milk for three times.
6. Samples are washed with PBS containing 5 % skim milk for three times. The slides are then incubated with FITC or carbocyanine (Cy) 3-conjugated 2nd Abs (dilution 1:100) at room temperature for 30 min. Then, the nucleus is labeled with DAPI for 10 min followed by washing with PBS for three times.
7. Stained samples are observed with a fluorescence microscope, and the percentage of positive cells are determined.
   The marker specific for eosinophils (EPO and MBP) increases with days of suspension culture, while the marker specific for basophil granulocytes (2D7 and pro-MBP) decreases (Fig. 4.2d).

4.3.9   The Functional Assay of hPSC-Eosinophils

For migration assay, we use Transwell T0 microchemotaxis chamber, 24-well plate, and different concentration of cytokines to detect the chemotactic migration of hPSC-derived eosinophils.

1. Growth factor dilutions (fMLP, IL-5, and eotaxin) in RPMI-1640/10 % FBS are filled into the lower wells and covered by the chemotaxis filter. Cells (1×10^5) in 50 μl RPMI-1640/10 % FBS are filled in the upper wells.
2. Plates are incubated at 37 °C in a humidified atmosphere containing 5 % CO_2 for 1 h. The remaining migrated cells on the lower side of the filter were counted.
3. The chemotactic migration of hPSC-derived eosinophils is detected in different concentration of fMLP (10 nM), eotaxin (50 nM), and IL-5 (1 ng/ml). The cell migration rate of hPSC-eosinophils is 42, 37.5, and 27 % when the concentration of fMLP, eotaxineotaxin, and IL-5 is 10 nM, 50 nM, and 1 ng/ml, respectively (Fig. 4.3a–c).

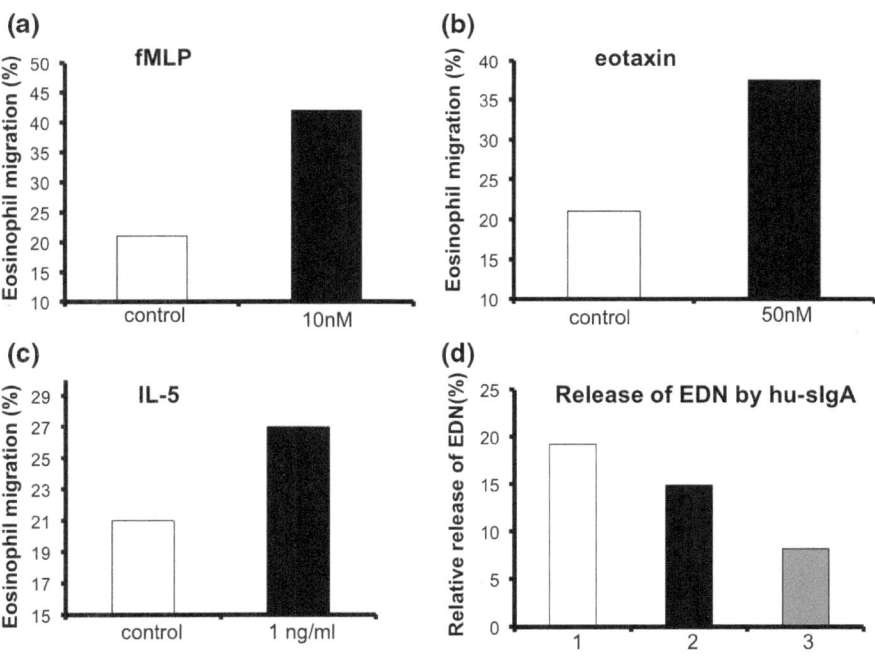

Fig. 4.3 Functional analysis of hESC-derived eosinophils. **a** Eosinophil migration upon stimulation of fMLP. **b** Eosinophil migration upon stimulation of eotaxin. **c** Eosinophil migration upon stimulation of IL-5. **d** EDN release upon stimulation of hu-sIgA. *d1* Eosinophils derived from peripheral blood. *d2* Eosinophils derived from hESCs. *d3* Eosinophils derived from cord blood CD34+ cells

The degranulation function of hPSC-derived eosinophils.

1. The concentration of EDN released by hPSC-derived eosinophils is measured through the stimulation of secreted IgA (hu-sIgA) according to the EDN ELISA kit.
2. In the stimulation of hu-sIgA, the concentration of EDN released by hPSC-derived eosinophils and cord blood CD34+ cell-derived and peripheral blood eosinophils are compared. We found that EDN released by hPSC-derived eosinophils is lower than that released by eosinophils from peripheral blood but higher than that released by eosinophils from cord blood CD34+ cell-derived eosinophils (Fig. 4.3d).

Notes:

1. All the materials which contact cells should be sterile.
2. Transferrin should be used within 1 week at 4 °C.
3. Matrigel should be thawed on ice to maintain liquid state for at least 1 h before coating; otherwise, the matrigel will turn to gel form at room temperature.
4. The method of introduction of eosinophils from hESCs can also be efficiently applied to hiPSCs. However, differences of production and purity could be observed among various hPSC lines.
5. The undifferentiated hPSC colonies should not be broken into single cells considering the recovering of hPSCs in cocultures.
6. IL-5 is an important cytokine for the maturation of hPSC-derived eosinophils, but not definitively needed in the early stage development.
7. All medium should be brought to room temperature prior to use.
8. Siglec-8 is an important surface marker for mature hPSC-derived eosinophils, so it is important to be used in FACS analysis.
9. Considering the degradation of RNA and the difficulty in abstracting RNA from mature eosinophils, it is recommended to use fresh cells and extract the RNA as soon as possible. All the procedure should be done on ice.

4.4 Discussion and Future Prospectives

In healthy individuals, eosinophils are often less in numbers in peripheral blood cells [29, 30]. The introduction of mature eosinophils from hPSCs provides us with unlimited supply of eosinophils for functional studies in vitro [8]. The hPSC-derived mature eosinophils possess similar properties to PB-derived eosinophils in migratory and phagocytic activities [8]. However, the expansion of eosinophils seemed to be relatively low in feeder-free conditions. We here introduce an efficient method to induce population of functionally mature eosinophils from hPSCs based on a coculture system. In our system, hPSC-derived eosinophils could be robustly generated with high purity (>90 %) in suspension culture within 3 weeks. Usually, $300 \sim 400 \times 10^6$ eosinophils could be generated from per 10^6 hPSCs. The hPSC-derived eosinophils not only phenotypically mimic the mature eosinophils from CB CD34+ cell-derived ones, but also exert mature functions as migration and release of EDN upon certain specific stimulations. This demonstrated that these hPSC-derived eosinophils from our coculture system in vitro mostly mimic the developmental and differentiational events to their counterparts in vivo. Our method may provide an excellent model to track the stepwise maturation of human eosinophils and evaluate the wide spectrum of eosinophil functions.

The analysis of the differentiation process of eosinophils can provide helpful information for the elucidation of the pathogenesis of hematopoietic diseases that affect eosinophils and/or myeloid differentiation. Blood diseases associated with defects in the maturation and function of eosinophils include eosinophil dysregulation and inherited bone marrow failure syndromes. Eosinophil dysregulation includes allergies, drug reactions, helminth infections, Churg–Strauss syndrome, some malignancies, metabolic disorders, eosinophilic gastrointestinal disorders, and hypereosinophilic syndrome [31, 32]. Inherited bone marrow failure syndromes such as eosinophilic leukemia and malignant medullar disease are associated with various chromosomal abnormalities including rearrangement of chromosome 12 and 5. The occurrence of mutation of genes encoding IL-3, IL-5, and GM-CSF on chromosome 5 may be involved in such hematologic malignancy [33]. It was believed that eosinophils could be primary contributors to disease pathophysiology, so eosinophils may prove to be a therapeutic target in controlling eosinophil-associated disorders [34]. The critical and expanding need for eosinophil-targeted therapies is highlighted in both developing and developed countries [35, 36]. Eosinophil-targeted therapeutic agents that are aimed at blocking specific steps involved in eosinophil development, migration, and activation have recently entered clinical testing and have produced encouraging results and insights into the role of eosinophils [32]. The hPSC-derived eosinophils, especially from patient-tailored iPSCs, provide an ideal model to understand the eosinophil development which will aid the research in new therapeutic strategies for diseases characterized by eosinophil dysregulation.

A long pending issue for the early development of human eosinophils is its lineage specification. The evidence accumulated in our laboratory clearly shows that hPSC-derived eosinophils follow an eosinophil–basophil common pathway in the early developmental stage, supporting the similar finding by Owen's group some two and half decades ago [37]. Since our system provides step-by-step route for both eosinophil and basophil developments, it may serve as a tool to search the molecular regulation controlling early development and lineage specification for human eosinophils and basophils.

Future studies should be emphasized on the following directions: (1) the early development of human hPSC-derived eosinophils, especially their phenotypical markers and molecular controls; (2) the functional comparison of hPSC-derived mature eosinophils with their peripheral counterparts; (3) the regulation of early development of human eosinophils and basophils; (4) establishing and banking of patient-specific iPSCs with severe allergic diseases, especially related to eosinophil development and hypersensitivities; (5) disease-tailored analysis of patient-specific iPSC-derived eosinophils and mechanism controlling the diseased conditions; and (6) drug screening by using patient iPSC-derived eosinophils to develop novel molecule-targeting therapies.

References

1. Thomson JA, Itskovitz-Eldor J, Shapiro SS, Waknitz MA, Swiergiel JJ, Marshall VS, Jones JM. Embryonic stem cell lines derived from human blastocysts. Science. 1998;282:1145–7.
2. Ma F, Ebihara Y, Umeda K, Sakai H, Hanada S, Zhang H, Zaike Y, Tsuchida E, Nakahata T, Nakauchi H. Generation of functional erythrocytes from human embryonic stem cell-derived definitive hematopoiesis. Proc Natl Acad Sci. 2008;105:13087–92.
3. Lu S-J, Feng Q, Park JS, Vida L, Lee B-S, Strausbauch M, Wettstein PJ, Honig GR, Lanza R. Biologic properties and enucleation of red blood cells from human embryonic stem cells. Blood. 2008;112:4475–84.
4. Saeki K, Saeki K, Nakahara M, Matsuyama S, Nakamura N, Yogiashi Y, Yoneda A, Koyanagi M, Kondo Y, Yuo A. A feeder-free and efficient production of functional neutrophils from human embryonic stem cells. Stem Cells. 2009;27:59–67.
5. Yokoyama Y, Suzuki T, Sakata-Yanagimoto M, Kumano K, Higashi K, Takato T, Kurokawa M, Ogawa S, Chiba S. Derivation of functional mature neutrophils from human embryonic stem cells. Blood. 2009;113:6584–92.
6. Gaur M, Kamata T, Wang S, Moran B, Shattil SJ, Leavitt AD. Megakaryocytes derived from human embryonic stem cells: a genetically tractable system to study Megakaryocytopoiesis and integrin function. J Thromb Haemost. 2005;4:436–42.
7. Takayama N, Nishikii H, Usui J, Tsukui H, Sawaguchi A, Hiroyama T, Eto K, Nakauchi H. Generation of functional platelets from human embryonic stem cells in vitro via ES-sacs, VEGF-promoted structures that concentrate hematopoietic progenitors. Blood. 2008;111:5298–306.
8. Choi K-D, Vodyanik MA, Slukvin II. Generation of mature human myelomonocytic cells through expansion and differentiation of pluripotent stem cell–derived lin–CD34+ CD43 + CD45+ progenitors. J Clin Investig. 2009;119:2818–29.
9. Slukvin II, Vodyanik MA, Thomson JA, Gumenyuk ME, Choi K-D. Directed differentiation of human embryonic stem cells into functional dendritic cells through the myeloid pathway. J Immunol. 2006;176:2924–32.
10. Woll PS, Martin CH, Miller JS, Kaufman DS. Human embryonic stem cell-derived NK cells acquire functional receptors and cytolytic activity. J Immunol. 2005;175:5095–103.
11. Kovarova M, Latour AM, Chason KD, Tilley SL, Koller BH. Human embryonic stem cells: a source of mast cells for the study of allergic and inflammatory diseases. Blood. 2010;115:3695–703.
12. Vodyanik MA, Bork JA, Thomson JA, Slukvin II. Human embryonic stem cell–derived CD34 + cells: efficient production in the coculture with OP9 stromal cells and analysis of lymphohematopoietic potential. Blood. 2005;105:617–26.
13. Timmermans F, Velghe I, Vanwalleghem L, De Smedt M, Van Coppernolle S, Taghon T, Moore HD, Leclercq G, Langerak AW, Kerre T. Generation of T cells from human embryonic stem cell-derived hematopoietic zones. J Immunol. 2009;182:6879–88.

14. Giarratana M-C, Kobari L, Lapillonne H, Chalmers D, Kiger L, Cynober T, Marden MC, Wajcman H, Douay L. Ex vivo generation of fully mature human red blood cells from hematopoietic stem cells. Nat Biotechnol. 2004;23:69–74.
15. Olsen AL, Stachura DL, Weiss MJ. Designer blood: creating hematopoietic lineages from embryonic stem cells. Blood. 2006;107:1265–75.
16. Ng ES, Davis RP, Azzola L, Stanley EG, Elefanty AG. Forced aggregation of defined numbers of human embryonic stem cells into embryoid bodies fosters robust, reproducible hematopoietic differentiation. Blood. 2005;106:1601–3.
17. Wang ZZ, Au P, Chen T, Shao Y, Daheron LM, Bai H, Arzigian M, Fukumura D, Jain RK, Scadden DT. Endothelial cells derived from human embryonic stem cells form durable blood vessels in vivo. Nat Biotechnol. 2007;25:317–8.
18. Kaufman DS, Hanson ET, Lewis RL, Auerbach R, Thomson JA. Hematopoietic colony-forming cells derived from human embryonic stem cells. Proc Natl Acad Sci. 2001;98:10716–21.
19. Qiu C, Hanson E, Olivier E, Inada M, Kaufman DS, Gupta S, Bouhassira EE. Differentiation of human embryonic stem cells into hematopoietic cells by coculture with human fetal liver cells recapitulates the globin switch that occurs early in development. Exp Hematol. 2005;33:1450–8.
20. Ledran MH, Krassowska A, Armstrong L, Dimmick I, Renström J, Lang R, Yung S, Santibanez-Coref M, Dzierzak E, Stojkovic M. Efficient hematopoietic differentiation of human embryonic stem cells on stromal cells derived from hematopoietic niches. Cell Stem Cell. 2008;3:85–98.
21. Yu J, Vodyanik MA, He P, Slukvin II, Thomson JA. Human embryonic stem cells reprogram myeloid precursors following cell–cell fusion. Stem Cells. 2006;24:168–76.
22. Matsuoka S, Tsuji K, Hisakawa H, Xu M-J, Ebihara Y, Ishii T, Sugiyama D, Manabe A, Tanaka R, Ikeda Y. Generation of definitive hematopoietic stem cells from murine early yolk sac and paraaortic splanchnopleures by aorta-gonad-mesonephros region–derived stromal cells. Blood. 2001;98:6–12.
23. Ma F, Kambe N, Wang D, Shinoda G, Fujino H, Umeda K, Fujisawa A, Ma L, Suemori H, Nakatsuji N. Direct development of functionally mature tryptase/chymase double-positive connective tissue-type mast cells from primate embryonic stem cells. Stem Cells. 2008;26:706–14.
24. Lee H, Shamy GA, Elkabetz Y, Schofield CM, Harrsion NL, Panagiotakos G, Socci ND, Tabar V, Studer L. Directed differentiation and transplantation of human embryonic stem cell-derived motoneurons. Stem Cells. 2007;25:1931–9.
25. Tian X, Woll PS, Morris JK, Linehan JL, Kaufman DS. Hematopoietic engraftment of human embryonic stem cell-derived cells is regulated by recipient innate immunity. Stem Cells. 2006;24:1370–80.
26. Ma F, Wang D, Hanada S, Ebihara Y, Kawasaki H, Zaike Y, Heike T, Nakahata T, Tsuji K. Novel method for efficient production of multipotential hematopoietic progenitors from human embryonic stem cells. Int J Hematol. 2007;85:371–9.
27. Ma F, Gu Y, Nishihama N, Yang W, Yasuhiro E, Tsuji K. Differentiation of human embryonic and induced pluripotent stem cells into blood cells in coculture with murine stromal cells. In: Ye K, Jin S, editors. Human embryonic and induced pluripotent stem cells (pp. 321–335). Totowa: Humana Press (2012). doi:10.1007/978-1-61779-267-0_23. ISBN 978-1-61779-266-3 (Published by Springer Science + Business Modia, LLC 2011).
28. Xu MJ, Tsuji K, Ueda T, Mukouyama YS, Hara T, Yang FC, Ebihara Y, Matsuoka S, Manabe A, Kikuchi A, Ito M, Miyajima A, Nakahata T. Stimulation of mouse and human primitive hematopoiesis by murine embryonic aorta-gonad-mesonephros-derived stromal cell lines. Blood. 1998;92:2032–40.
29. Shamri R, Xenakis JJ, Spencer LA. Eosinophils in innate immunity: an evolving story. Cell Tissue Res. 2011;343:57–83.
30. Lacy P, Rosenberg HF, Walsh GM. Eosinophil overview: structure, biological properties, and key functions. In Eosinophils. Berlin: Springer; 2014. pp 1–12.

31. Rosenberg HF, Dyer KD, Foster PS. Eosinophils: changing perspectives in health and disease. Nat Rev Immunol. 2012;13:9–22.
32. Fulkerson PC, Rothenberg ME. Targeting eosinophils in allergy, inflammation and beyond. Nat Rev Drug Discovery. 2013;12:117–29.
33. Crescenzi B, Chase A, La Starza R, Beacci D, Rosti V, Galli A, Specchia G, Martelli M, Vandenberghe P, Cools J. FIP1L1-PDGFRA in chronic eosinophilic leukemia and BCR-ABL1 in chronic myeloid leukemia affect different leukemic cells. Leukemia. 2007;21:397–402.
34. Bochner BS, Gleich GJ. What targeting eosinophils has taught us about their role in diseases. J Allergy Clin Immunol. 2010;126:16–25.
35. Hruz P, Straumann A, Bussmann C, Heer P, Simon H-U, Zwahlen M, Beglinger C, Schoepfer AM. Escalating incidence of eosinophilic esophagitis: a 20-year prospective, population-based study in Olten County, Switzerland. J Allergy Clin Immunol. 2011;128 (1349–1350):e1345.
36. Bohm M, Malik Z, Sebastiano C, Thomas R, Gaughan J, Kelsen S, Richter JE. Mucosal eosinophilia: prevalence and racial/ethnic differences in symptoms and endoscopic findings in adults over 10 years in an urban hospital. J Clin Gastroenterol. 2012;46:567–74.
37. Boyce JA, Friend D, Matsumoto R, Austen KF, Owen WF. Differentiation in vitro of hybrid eosinophil/basophil granulocytes: autocrine function of an eosinophil developmental intermediate. J Exp Med. 1995;182:49–57.

Chapter 5
Human Pluripotent Stem Cells as a Renewable Source of Natural Killer Cells

David L. Hermanson, Zhenya Ni and Dan S. Kaufman

Abstract Human pluripotent stems cells provide an ideal source for the study of hematopoietic differentiation. Natural killer (NK) cells are lymphocytes that play a key role in innate immunity against viral infections as well as malignancies. The development and differentiation of NK cells have been an area of increasing research interest due to their clinical utility in treating multiple types of cancer and potentially infectious disease. Our initial studies to derive NK cells from human embryonic stem cells (hESCs) and induced pluripotent stem cells (iPSCs) used a stromal cell co-culture method with relatively poor-defined conditions. Subsequent studies have utilized a stroma-free embryoid body (EB) method to generate hemato-endothelial precursor cells followed by in vitro NK cell differentiation in defined conditions. Further expansion of these hESC- and iPSC-derived NK cells can be done through the use of interleukin (IL)-21 expressing artificial antigen-presenting cells (aAPCs). Combining these methods, we can efficiently generate enough NK cells required for clinical therapies from a small number of undifferentiated human pluripotent stem cells. These methods enable hESCs and iPSCs to be used to produce an essentially unlimited amount of homogenous NK cells that can be used as a standardized, off-the-shelf immunotherapy for the treatment of refractory cancers and other diseases.

Keywords Embryonic stem cells · Induced pluripotent stem cells · Embryoid body · Hematopoietic progenitors · Natural killer cells

D.L. Hermanson · Z. Ni · D.S. Kaufman (✉)
Department of Medicine and Stem Cell Institute, University of Minnesota,
420 Delaware St. SE, MMC 480, Minneapolis, MN 55455, USA
e-mail: kaufm020@umn.edu

© The Author(s) 2015
T. Cheng (ed.), *Hematopoietic Differentiation of Human Pluripotent Stem Cells*,
SpringerBriefs in Stem Cells, DOI 10.1007/978-94-017-7312-6_5

5.1 Introduction

Natural killer (NK) cells represent an important lymphocyte population essential to the innate immune system with a role in protecting the host from both viral infections and malignancies. NK cells isolated from peripheral blood (PB-NK cells) have shown valuable use in the clinic for the treatment of acute myeloid leukemia (AML) as well as other malignancies [1–3]. However, the need for donors and cell processing may limit the use of PB-NK cells. Additionally, PB-NK cells are a heterogeneous cell population, with often less than 50 % NK cells in the product given to patients and will vary from each donor. A method to produce an "off-the-shelf" NK cell product from human pluripotent stem cells could overcome these hurdles. In vitro methods for the differentiation of NK cells from CD34+ cells isolated from human bone marrow or umbilical cord blood (UCB) have previously been established [4–7]. In these studies, NK cells were produced by culture of CD34+ cells with stem cell factor (SCF), IL-7, and Flt3 ligand (Flt3L). These conditions lead to cells that express IL-15 receptor, which has been shown to be critical for the formation of NK cells. Final NK cell differentiation is then driven by the addition of IL-15. Additional studies have aimed to improve the NK cell production from primary CD34+ cells using IL-21 signaling to mediate expansion of NK cells and stimulate expression of CD16 [8, 9], an Fc receptor typically expressed by NK cells and crucial for their ability to mediate antibody-dependent cell-mediated cytotoxicity.

Human pluripotent stem cells represent an ideal starting population for the development of multiple cell types, including NK cells. Human pluripotent stem cells include embryonic stem cells (ESCs) and induced pluripotent stem cells (iPSCs), both of which represent ideal starting populations to further study the development of NK cells in vitro. Furthermore, NK cells from hESC/iPSC provide a standardized cell-based treatment for malignant hematopoietic and solid tumors [10, 11]. However, the advancement of these treatments requires a reliable procedure for the production of large numbers of NK cells from hESC/iPSCs.

This chapter describes a method for the efficient generation of NK cells from hESC/iPSC. As previously described, both hESC- and iPSC-derived NK cells (hESC-NKs, iPSC-NKs) express activating and inhibitory receptors similar to NK cells isolated from adult peripheral blood [12–14]. The hESC-derived NK cells are also highly efficient at direct cell-mediated cytotoxicity and antibody-dependent cell-mediated cytotoxicity, as well as production of cytokines such as interferon-γ. Stromal cell co-culture or stroma-free systems can be used to generate hemato-poietic progenitor cells, both of which are capable of developing into functional NK cells. For ease of clinical translation, we have recently adopted a "spin EB" protocol [15, 16] to provide a system for more consistent hematopoietic cell differentiation without the use of xenogeneic stromal cells (Fig. 5.1). Following the formation of the hemato-endothelial precursor cells through the "spin EB" process, NK cells are differentiated from the embryoid bodies (EBs) using procedures similar to those used for CD34+ cells isolated from peripheral blood, or cord blood. Again, this NK

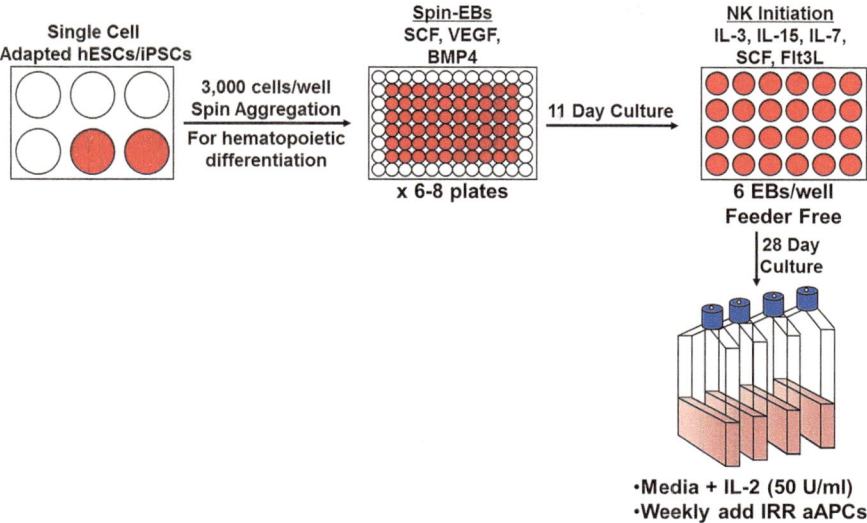

Fig. 5.1 Schematic diagram of two-step hematopoietic and NK cell differentiation from human embryonic stem cells or induced pluripotent stem cells. In brief, undifferentiated hESC/iPSCs are induced to differentiate into hematopoietic progenitor cells by spin EB formation. The hematopoietic progenitor cells can be characterized based on expression of specific cell surface markers (typically CD34 and CD45). Spin EBs can be directly transferred into NK cell development conditions with defined media plus cytokines. After 3–5 weeks, mature and functional NK cells develop. NK cells can be further expanded using aAPCs and IL-2

cell differentiation step can be done with support from stromal cell co-culture or can be done without the aid of stromal cells for ease of clinical translation.

5.2 Materials

5.2.1 Cell Lines

1. H1 and H9 line human ES cells (WiCell, Madison, WI). Multiple iPS cell lines have been generated and used in our laboratory for the expression of the four transcription factors Oct4, Sox2, c-Myc, and Klf4 in CD34$^+$ UCB cells, human fibroblasts, or human peripheral blood cells. hESCs/iPSCs are maintained as undifferentiated cells as previously described [17] (Note 1).
2. EL08-1D2 stromal cells (Kindly provided by Dr. Rob Oostendorp, Technical University, Munich, Germany) [18, 19]. These cells are maintained in 50 % Myelocult M5300 (StemCell Technologies, cat. no. 05350), 35 % Alpha Minimum Essential Media (Invitrogen, cat. no. 12571-063), 15 % FBS (StemCell Technologies, cat. no. 06500), 1 % Glutamax 1 (100×, Invitrogen,

cat. no. 35050-061), 0.1 mM β-mercaptoethanol, 10^{-6} M hydrocortisone (StemCell Technologies, cat. no. 07904), and 1 % penicillin/streptomycin.

3. K562 cells expressing membrane-bound IL-21 (Clone 9. mbIL-21) (Kindly provided by Drs. Dean Lee and Lawrence Cooper, MD Anderson Cancer Center, Houston, TX) [20]. These cells are maintained in RPMI-1640 plus L-glutamine, 10 % FBS, and 1 % penicillin/streptomycin.

5.2.2 Hematopoietic Differentiation of hES/iPS Cells by Spin EBs

1. Spin EB differentiation medium (BPEL media) [15]: 43 % Iscove's Modified Dulbecco's Medium (IMDM, Thermo, cat. no. SH30228.01), 43 % F-12 Nutrient Mixture w/Glutamax I (Invitrogen, cat. no. 3176503), 0.25 % deionized bovine serum albumin (BSA, Sigma-Aldrich, cat. no. A3311), 0.25 % poly (vinyl alcohol) (Sigma-Aldrich, cat. no. P8136), 0.1 ug/mL linoleic acid (Sigma-Aldrich, cat. no. L1012), 0.1 ug/mL linolenic acid (Sigma-Aldrich, cat. no. L2376), 1:500 Synthechol 500× solution (Sigma-Aldrich, cat. no. S5442), 450 uM α-monothioglycerol (α-MTG) (Sigma-Aldrich, cat. no. M6145), 5 % protein-free hybridoma mix II (Invitrogen, cat. no. 12040077), 50 ug/mL ascorbic acid 2-phosphate (Sigma-Aldrich, cat. no. A8960), 2 mM Glutamax I (Invitrogen, cat. no. 35050061), 1 % Insulin, Transferrin, and Selenium 100× solution (ITS) (Invitrogen, cat. no. 41400-045), 1 % penicillin/streptomycin plus 40 ug/mL recombinant human SCF (PeproTech, cat. no. 300-07), 20 ug/mL BMP4 (R&D systems, cat. no. 314-BP), and 20 ug/mLVEGF (R&D systems, cat. no. 293-VE).
2. TrypLE Select (Gibco/Invitrogen, cat. no. 12563-011).
3. Dulbecco's phosphate buffer saline (HyClone, cat. No. SH30028.02).
4. 96-well round bottom plates (NUNC, cat. no. 262162 with lids cat. no. 264122).

5.2.3 Natural Killer Cell Differentiation from Differentiated Spin EBs

1. NK cell differentiation medium: 56.6 % DMEM-high glucose, 28.3 % HAMS/F12 (Invitrogen, cat. no. 11765-064), 15 % heat-inactivated human AB serum (Valley Biomedical, cat. no. HP1022 HI), 2 mM L-glutamine, 1 uM β-mercaptoethanol, 5 ng/mL sodium selenite (Sigma-Aldrich, cat. no. S5261), 50 uM ethanolamine (MP Biomedicals, cat. no. 194658), 20 mg/L ascorbic acid (Sigma-Aldrich, cat. no. A-5960), 1 % P/S, 5 ng/mL IL-3 (PeproTech, cat. no. 200-03), 20 ng/mL SCF, 20 ng/mL IL-7 (PeproTech, cat. no.), 10 ng/mL IL-15

(PeproTech, cat. no. 200-15), and 10 ng/mL Flt3 ligand (Flt3L) (PeproTech, cat. no. 300-19). Store at 4 °C in the dark.
2. 24-well tissue culture plates (NUNC Brand Products, Nalgene Nunc; cat. no. 142475).

5.2.4 Natural Killer Cell Expansion

1. NK cell expansion medium: RPMI-1640, 10 % FBS, 2 mM L-glutamine, 1 % P/S, and 50 U/mL IL-2 added just prior to NK cells.

5.3 Methods

5.3.1 TrypLE Adaptation of hESC/iPSCs

Prior to the generation of spin EB, undifferentiated hESCs/iPSCs must be TrypLE adapted, a method previously described [15, 21]. This process allows for the hESC/iPSC to be made more efficiently into single-cell suspensions. Prior to adaptation, carefully remove differentiated hES/iPS cells and colonies because differentiated cells can easily dominate the cultures. **It is very important to make sure your starting population does not contain differentiated cells.** Cells can usually be used to set up spin EBs after 12–15 passages in TrypLE.

1. The day prior to passing hESC/iPSCs, seed gelatin-coated 6-well plates with low-density MEFs. Low-density MEFs are plated at ½ the amount of what would be used to support hESC/iPSC cultures. We use 90,000 cells/well of a 6-well plate as low density.
2. After making sure the starting population is free of differentiated colonies, aspirate media and pass hESC/iPSCs by incubating cells for 5 min at 37 °C in pre-warmed TrypLE.
3. Collect cells by gentle pipetting and dilute using equal parts of hESC/iPSC media and DPBS.
4. Centrifuge cells and aspirate media.
5. Wash cells once by resuspending cell pellet in media and DPBS and then centrifuge again.
6. Plate hESC/iPSCs 1:1 onto low-density MEFs plated the day prior to passing.
7. Replace media daily and repeat procedure when cells become confluent. After ∼5 passages, split cells 1:2 or 1:3 and eventually cells can be split 1:6 twice a week.

5.3.2 Generation of Hematopoietic Progenitor Cells from hES/iPS Cells by Spin EB Formation

Many studies have investigated hematopoietic differentiation by embryoid body (EB) formation. Advantages of this method compared to stromal cell co-culture include defined culture conditions and higher efficiency of hematopoietic progenitor cell generation. However, the EB system can be compared with the co-culture system [22, 23]. The spin EB approach developed by Elefanty's group [15, 21] allows for a more consistent generation of hematopoietic progenitors.

1. Two days prior to setting up spin EB differentiation, pass 200,000–250,000 TrypLE-adapted hESC/iPSCs onto fresh MEFs. They should be 70–80 % confluent on the day of spin EBs setup. Generally 2–3 wells of a 6-well plate are sufficient for 5 plates of spin EB.

2. To prepare for Spin EB plating, pipet 150 μL sterile water into the 36 outer wells of each 96-well plate to minimize loss of well volume to evaporation.

3. Aspirate culture media from ES/iPS cells and add 1 ml pre-warmed TrypLE to each well. Place plates in cell culture incubator (5 % CO2, 37 °C) for 4–5 min until hESC/iPSC cells start to come off the plate.

4. Collect dissociated cells in a conical tube and pipet up and down to break up clumps. Dilute TrypLE with 1 volume BPEL media and at least 1 volume DPBS. Centrifuge cells, remove supernatant, and resuspend the cells in 5 ml BPEL media plus 5 ml DPBS. Centrifuge cells again.

5. Remove supernatant and resuspend cells in 5 ml BPEL media. Pass cells through 70-μm filter into a fresh conical tube in order to remove clumps. Count filtered cells and aliquot the appropriate number of cells to be used for plating. Typically, cells are seeded at 3000 cells/well in the 96-well plates, though this density can be varied if desired. Centrifuge and resuspend hESC/iPSCs in BPEL containing cytokines at a concentration of 3×10^4 cells/ml.

6. Pipet 100 μl of the cell suspension into each of the inner 60 wells of the prepared 96-well plates with 150 μl of water in outer wells using a multi-channel pipet. Centrifuge 96-well plates at 480 g, 8 °C for 4 min, and incubate the plates at 37 °C, 5 % CO_2 for 8–11 days (Fig. 5.2). We have found with our hESC/iPSCs that 11 days is optimal for the formation of hematopoietic progenitors defined by the presence of $CD34^+CD45^+$ cells. It is important not to disturb the plates during the first 3 days of differentiation while the spin EBs are forming (Note 2). Under optimal conditions, the percentage of $CD34^+$ cells can be approximately 40–60 % and percentage of $CD34^+CD45^+$ can be up to 20–40 %. While there is variability between different hESC and iPSC lines, some lines can be used up to passage 40–50 after TrypLE adaptation to make spin EBs.

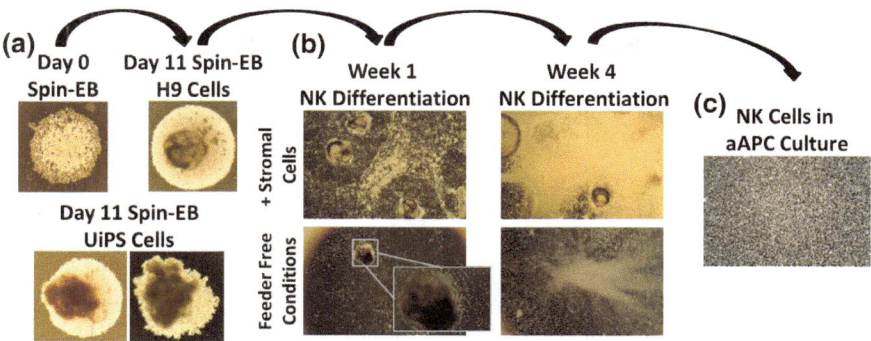

Fig. 5.2 Phase microscope images of developing NK cells from human pluripotent stem cells. **a** Images taken of spin EBs at day 0 and day 11, immediately before transfer to NK cell differentiation conditions. Spin EBs were made using 3000 cells/well and cultured in SCF, BMP-4, and VEGF. **b** Images of NK cell differentiation with or without stromal cells. In the enlarged image, non-adherent hematopoietic cells can be seen being produced from the EB population, which is still intact after transfer. In these conditions, the spin EBs also produce adherent endothelial cells and mesenchymal cells. **c** Images of NK cells being grown in aAPC expansion culture. At this point, the cells are a homogenous NK cell population

5.3.3   Natural Killer Cell Differentiation with or without Stromal Cells

Transfer of spin EBs can be done into 24-well plates with either stromal cells or in a feeder-free method. We have observed that with stromal cells we get slightly higher yields of NK cells, but the feeder-free method is better suited for clinical translation [23].

1. If stromal cell co-culture is desired: The EL08-1D2 (note 3) feeder plates are usually prepared the day before transferring spin EBs. Dissociate cells from flask with 0.05 % trypsin. Collect cells by centrifugation and resuspend in 80 % fresh medium +20 % old medium to $1.2–1.25 \times 10^5$/mL. Cell are plated onto 0.1 % gelatin-coated 24-well plates at 1 mL/well and will grow to 85–90 % confluency at 33 °C, 5 % CO_2 by the following day. The plates are then X-ray irradiated with 3000 rad, and the feeders are ready for use.

2. Spin EBs (without sorting) are directly transferred into 24-well plates on day 11, with or without stromal cells. Using a multi-channel pipet, carefully remove ½ the volume from the 96-well plates prior to transferring. We transfer 6–8 EBs into each well of the 24-well plate using the multi-channel pipet where two tips fit into each well. Each well then receives 0.5–0.7 mL of NK differentiation medium containing all of the cytokines (note 4).

3. Half-medium changes are done every 5–7 days. Following the first week of the NK cell differentiation, IL-3 is no longer added to the media.

4. Continue half-medium changes for 4–5 weeks to achieve mature NK cells (Fig. 5.2) (note 5). Cells can then be collected by passing them through a 70-μm filter to remove any clumps.

5. Mature CD45$^+$CD56$^+$ NK cells can be phenotyped by flow cytometry. In vitro function of hES/iPS cell-derived NK cells can be analyzed by the measurement of direct cytolytic activity tumor cells (such as K562) by a standard ^{51}Cr-release assay or immunological assays for cytotoxic granule or cytokine release [13, 14]. NK cells developed under feeder-free condition also show a mature NK cell phenotype and cytotoxicity.

5.3.4   Clinical Scale Expansion of hESC-/iPSC-Derived NK Cells for Immunotherapy

From two 24-well plates, we regularly able to produce 3–10 × 10^6 NK cells. In order to further expand the NK cells, artificial antigen-presenting cells (aAPCs) are used to generate >10^9 NK cells (Fig. 5.3) [20, 23].

1. In order to prepare for expansion of NK cells, aAPCs have to be cultured and irradiated. We utilize mbIL-21 K562 s that receive lethal irradiation with 10,000 rads. These can be prepared on the day of expansion or frozen stocks can be made ahead of time.

2. Following NK cell differentiation, the cells are collected by passing them through a 70-μm filter. The cells are then centrifuged and resuspended at a density of 2 × 10^5 cells/mL in NK expansion medium containing 50 U/mL IL-2 added freshly.

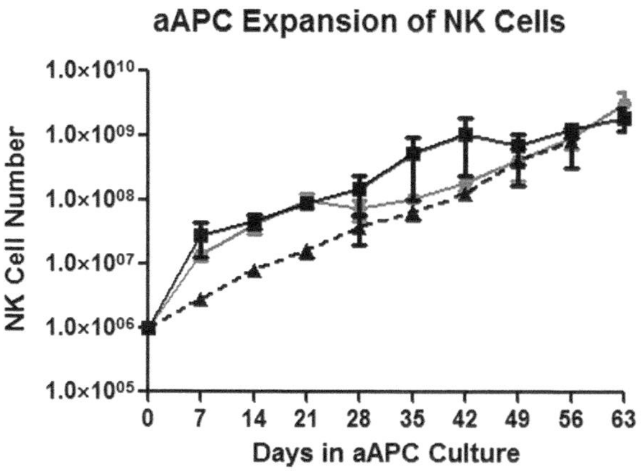

Fig. 5.3 Expansion data for the growth of NK cells using aAPCs. H9 ESC-NK cells grown in EL08-ID2 stromal cells (*black line*), H9 ESC-NK cells grown in feeder-free conditions (*gray line*), and iPSC-NK cells grown in feeder-free conditions (*black dashed line*) were transferred from NK cell differentiation conditions and placed in aAPCs for expansion. Cells were stimulated with aAPCs weekly and expanded for 9 weeks. Cell number was normalized to 1 × 10^6 cells on D0. Each line represents mean of 2 independent experiments

3. The NK cells are then stimulated using aAPCs every 7 days. Initially, we use 2:1 aAPC:NK cells for the first week and thereafter a ratio of 1:1. Media are changed every 3–4 days containing 50 U/mL of freshly added IL-2.
4. NK cells can be expanded for >60 days without a decrease in cell viability or cytolytic activity.

5.4 Notes

1. We have successfully performed NK cell differentiation from multiple iPS cell lines derived from various starting cell populations.
2. EBs should start to form after 24-h incubation, but if hESC/iPSCs have not been well adapted by TrypLE (at least >10 passages), spin EBs may not be formed well. Also, BPEL components should be changed every 3–4 months with the exception of PVA and deionized BSA, which should be made every 4 weeks.
3. EL08-1D2 stromal cells are a temperature-sensitive cell line, which should be maintained at an incubator with 33 °C and 5 % CO_2. However, cells can be kept at 37 °C after inactivation. Additionally, other stromas such as OP9, OP9DL1, and OP9DL4 have been compared with EL08-1D2 for NK cell differentiation in our laboratory. OP9DL1 s were found to be just as efficient as EL08-1D2.
4. Poor NK cell differentiation from hematopoietic progenitors on EL08-1D2 stromal cell co-cultures. If these feeder cells have been passed in culture for more than two months (typically more than 10 passages), they may not support NK cell development effectively. The resource and different lots of cytokines also affect NK cell differentiation. Also make sure cytokines are stored at −20 ° C long term and no longer than a week at 4 °C. If necessary, use fresh stromal cells every week until mature NK cells development.
5. It usually takes longer (about 5 weeks) to derive NK cells from spin EBs than from enriched progenitors from stromal co-culture. It is likely that hematopoietic progenitor cells in day-11 spin EBs are still in earlier development status compared with those from day 21 M210-B4 co-cultures.

Acknowledgments Support for studies of NK cell development have come from National Institutes of Health/NHLBI, the University of Minnesota Masonic Cancer Center, the William Lawrence & Blanche Hughes Foundation, the State of Minnesota Partnership for Biotechnology and Medical Genomics, the Minnesota Ovarian Cancer Alliance, and the International Clinical Research Center.

References

1. Ljunggren HG, Malmberg KJ. Prospects for the use of NK cells in immunotherapy of human cancer. Nat Rev Immunol. 2007;7:329–39.
2. Miller JS, Soignier Y, Panoskaltsis-Mortari A, et al. Successful adoptive transfer and in vivo expansion of human haploidentical NK cells in patients with cancer. Blood. 2005;105:3051–7.

3. Geller MA, Cooley S, Judson PL, et al. A phase II study of allogeneic natural killer cell therapy to treat patients with recurrent ovarian and breast cancer. Cytotherapy. 2011;13:98–107.

4. Miller JS, Alley KA, McGlave P. Differentiation of natural killer (NK) cells from human primitive marrow progenitors in a stroma-based long-term culture system: identification of a CD34 + 7 + NK progenitor. Blood. 1994;83:2594–601.

5. Miller JS, McCullar V, Verfaillie CM. Ex vivo culture of CD34 +/Lin-/DR- cells in stroma-derived soluble factors, interleukin-3, and macrophage inflammatory protein-1alpha maintains not only myeloid but also lymphoid progenitors in a novel switch culture assay. Blood. 1998;91:4516–22.

6. Silva MR, Kessler S, Ascensao JL. Hematopoietic origin of human natural killer (NK) cells: generation from immature progenitors. Pathobiology. 1993;61:247–55.

7. Mrózek E, Anderson P, Caligiuri MA. Role of interleukin-15 in the development of human CD56 + natural killer cells from CD34 + hematopoietic progenitor cells. Blood. 1996;87:2632–40.

8. Sivori S, Cantoni C, Parolini S, et al. IL-21 induces both rapid maturation of human CD34 + cell precursors towards NK cells and acquisition of surface killer Ig-like receptors. Eur J Immunol. 2003;33:3439–47.

9. Perez SA, Mahaira LG, Sotiropoulou PA, et al. Effect of IL-21 on NK cells derived from different umbilical cord blood populations. Int Immunol. 2006;18:49–58.

10. Wilber A, Linehan JL, Tian X, et al. Efficient and stable transgene expression in human embryonic stem cells using transposon-mediated gene transfer. Stem Cells. 2007;25:2919–27.

11. Giudice A, Trounson A. Genetic modification of human embryonic stem cells for derivation of target cells. Cell Stem Cell. 2008;2:422–33.

12. Woll PS, Martin CH, Miller JS, Kaufman DS. Human embryonic stem cell-derived NK cells acquire functional receptors and cytolytic activity. J Immunol. 2005;175:5095–103.

13. Ni Z, Knorr DA, Clouser CL, et al. Human pluripotent stem cells produce natural killer cells that mediate anti-HIV-1 activity by utilizing diverse cellular mechanisms. J Virol. 2011;85:43–50.

14. Woll PS, Grzywacz B, Tian X, et al. Human embryonic stem cells differentiate into a homogeneous population of natural killer cells with potent in vivo antitumor activity. Blood. 2009;113:6094–101.

15. Ng ES, Davis R, Stanley EG, Elefanty AG. A protocol describing the use of a recombinant protein-based, animal product-free medium (APEL) for human embryonic stem cell differentiation as spin embryoid bodies. Nat Protoc. 2008;3:768–76.

16. Ng ES, Davis RP, Azzola L, Stanley EG, Elefanty AG. Forced aggregation of defined numbers of human embryonic stem cells into embryoid bodies fosters robust, reproducible hematopoietic differentiation. Blood. 2005;106:1601–3.

17. Kaufman DS, Hanson ET, Lewis RL, Auerbach R, Thomson JA. Hematopoietic colony-forming cells derived from human embryonic stem cells. Proc Natl Acad Sci USA. 2001;98:10716–21.

18. Oostendorp RA, Robin C, Steinhoff C, et al. Long-term maintenance of hematopoietic stem cells does not require contact with embryo-derived stromal cells in cocultures. Stem Cells. 2005;23:842–51.

19. Ledran MH, Krassowska A, Armstrong L, et al. Efficient hematopoietic differentiation of human embryonic stem cells on stromal cells derived from hematopoietic niches. Cell Stem Cell. 2008;3:85–98.

20. Denman CJ, Senyukov VV, Somanchi SS, et al. Membrane-bound IL-21 promotes sustained ex vivo proliferation of human natural killer cells. PLoS ONE. 2012;7:e30264.

21. Ng ES, Davis RP, Hatzistavrou T, Stanley EG, Elefanty AG. Directed differentiation of human embryonic stem cells as spin embryoid bodies and a description of the hematopoietic blast colony forming assay. Curr Protoc Stem Cell Biol Chapter. 2008;1: Unit 1D 3.
22. Hexum MK, Tian X, Kaufman DS. In vivo evaluation of putative hematopoietic stem cells derived from human pluripotent stem cells. Methods Mol Biol. 2011;767:433–47.
23. Knorr DA, Ni Z, Hermanson D, et al. Clinical-scale derivation of natural killer cells from human pluripotent stem cells for cancer therapy. Stem Cells Transl Med. 2013;2:274–83.

Chapter 6
Generation of T-Lineage Cells from iPS Cells and Its Application

Haruka Wada, Muhammad Baghdadi and Ken-ichiro Seino

Abstract The development of cell reprogramming technology will alter many clinical situations, including immunotherapy. One instance is donor lymphocyte infusion (DLI) therapy. DLI is an effective therapy for lymphoma or leukemia patient for the inhibition of relapses after bone marrow transplantation (BMT). However, DLI treatment depends on the availability of T lymphocytes isolated from a donor for the generation of lymphocytes. The application of cell reprogramming technology facilitates the generation of lymphocytes without donor apheresis, offering a sustainable and repeatable DLI therapy. In this chapter, we show an instance of this new age immunotherapy that utilizes cell reprogramming technology.

Keywords Cell reprogramming technology · T cells · Donor lymphocyte infusion therapy · Differentiation · iPS cells

6.1 Introduction

The development of iPS cell technology [1] has opened a new world of regenerative medicine and will have great impact on future medical applications including cancer immunotherapy. Cancer immunotherapy strategies utilize various immune cells to initiate or amplify anti-tumor responses leading to tumor rejection and destruction. These therapeutic interventions using immune cells are considered now as one of the most promising strategies for cancer treatment in the future [2]. Importantly, the therapeutic effects of current cancer treatments including surgery, chemotherapy, and radiation were found to be significantly enhanced when combined with immunotherapy approaches [3]. However, one obstacle in immune

H. Wada · M. Baghdadi · K. Seino (✉)
Institute for Genetic Medicine, Hokkaido University,
Kita-15 Nishi-7, Sapporo 060-0815, Japan
e-mail: seino@igm.hokudai.ac.jp

© The Author(s) 2015
T. Cheng (ed.), *Hematopoietic Differentiation of Human Pluripotent Stem Cells*,
SpringerBriefs in Stem Cells, DOI 10.1007/978-94-017-7312-6_6

81

cell-based therapeutic strategies is the difficulty to obtain enough numbers of effector immune cells from the patient. The new iPS cell technology serves as a powerful tool to overcome such difficulties and further improve and accelerates current treatments. Such innovative strategies are eagerly anticipated at the present.

Here, we proposed a new strategy which combines between donor lymphocyte infusion (DLI) therapy [4] and cell reprogramming technology for the inhibition of relapses in leukemia model. Usually, DLI treatment needs donor apheresis to obtain lymphocytes in every treatment course; however, this novel strategy facilitates obtaining T cells for DLI therapy without donor apheresis. Furthermore, in this new DLI technology, a continuous DLI treatment can be done every time when needed.

We describe here a protocol used in DLI therapy in our laboratory. In this protocol, we first prepared iPS cells from donor bone marrow-derived cells. After examining the quality of iPS cells, T-lineage cells are differentiated from these iPS cells (iPS-T cells). iPS cells were trypsinized and disaggregated into a single-cell suspension. Single-cell suspension of iPS cells was put into leukemia inhibitory factor (LIF)-free media to form embryoid body-like sphere [5]. Differentiation of iPS cells was induced with a withdrawal of LIF from the culture in a non-treated plastic dish. About five days later, sphere formation was observed (Fig. 6.1a). Embryoid body-like sphere includes three types of germ layer cells. If the differentiation culture succeeded, ~ 30 % of generated cells express Fetal liver kinase-1 (Flk-1) (Fig. 6.1d), which is a mesoderm cell marker [6, 7]. Single-cell suspension containing mesoderm cells put on OP9/Delta like-1 (OP9/DL1), which enforces OP9 stromal cells to express Delta like-1 molecules, using appropriate cytokines such as Fms-related tyrosine kinase 3 (Flt3) ligand. These mesoderm cells will give rise to hematopoietic lineage cells [7]. OP9 cells support hematopoietic cell differentiation [8]. The generation of T-lineage cells from hematopoietic cells needs Delta like-1 or Delta like-4 molecules [9, 10]. These molecules are Notch ligands that have been shown to be essential for T-lineage cell differentiation [9, 10]. Several days later, immature hematopoietic cell cluster on stromal cells was observed by microscope (Fig. 6.1b) [5], and CD45 expression was confirmed by flow cytometry. Loosely attached hematopoietic cells were harvested by gentle pipetting and put on to new OP9/DL1 stromal cells. Lymphocyte-like expansions were observed at day 14 (Fig. 6.1c), and these cells contained immature T cells (Fig. 6.1d). Around 20 days later from first differentiation, iPS-T cells included populations of αβ-T cells, γδ-T cells, and NK cells (Fig. 6.1d). In terms of developmental stage of T cells, the majority of generated cells are CD4 and CD8 double-positive T cells, while the rest small population includes CD8 single-positive T cells (Fig. 6.1d). These iPS-T cells express various types of T cell receptors (TCRs) [5]. To determine whether TCRs expressed on these T cells were indeed functional, we evaluated IFN-γ production after TCR stimulation. To test this, iPS-T cells were stimulated in a plate-bound anti-CD3 antibody for 2 days, and IFN-γ production was evaluated by flow cytometry. Certain populations of iPS-T cells produced IFN-γ in response to TCR stimulation (Fig. 6.1e).

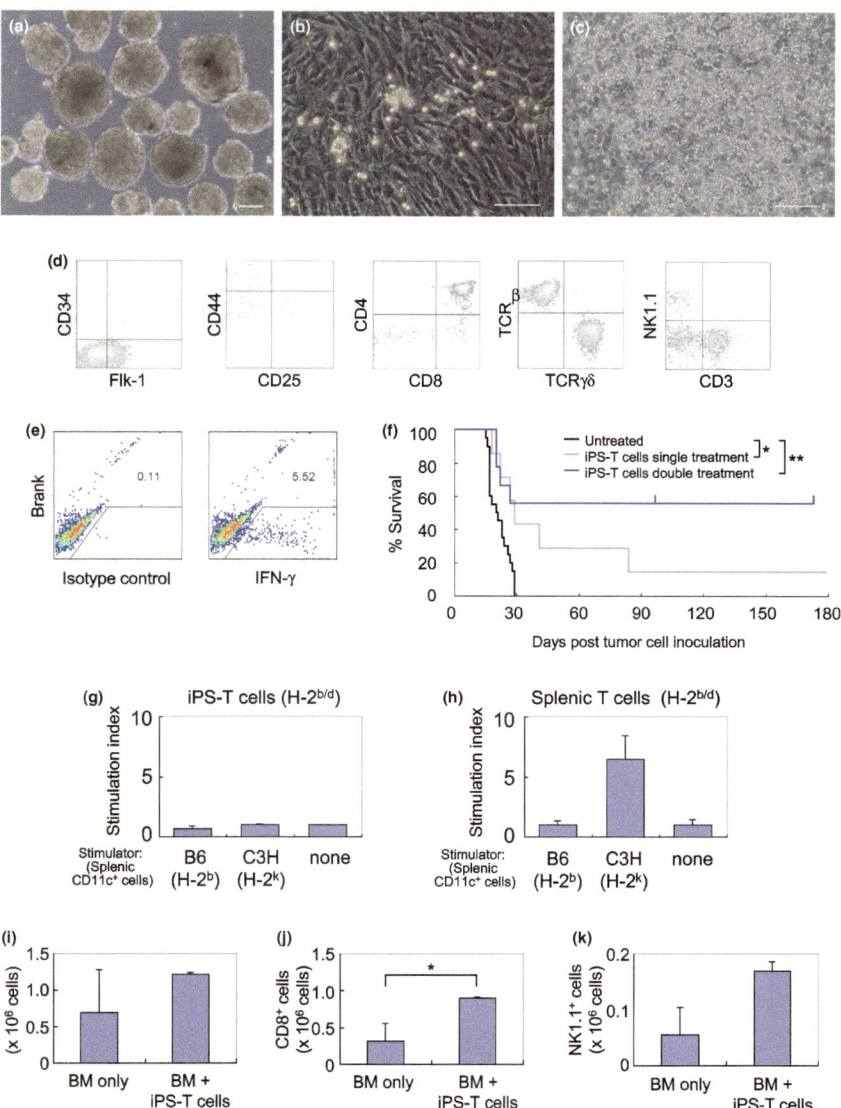

◀ **Fig. 6.1** Generation of T-lineage cells from iPS cells and its application. **a** Embryoid body-like sphere of iPS cells. iPS cells were cultured in OP9/DL1 media without LIF for 5 days. Bar: 100 μm. **b** Hematopoietic cell formation. The single-cell suspension of embryoid body-like spheres was cultured on OP9 cells for another 3 days in the presence of Flt3 ligand. Bar: 100 μm. **c** Lymphocyte-like proliferation. At day 14, appearance of cells co-cultured with OP9-DL1 in the presence of Flt3 ligand and IL-7. Scale bar: 100 μm. **d** Flow cytometry analysis. Expression of CD34 and Flk-1 (at day 5), CD44 and CD25 (at day 14), CD4 and CD8 (at day 20–26), TCRβ and TCRγδ (at day 20–26), and CD3 and NK1.1 (at day 20–26) in generated cells was analyzed. **e** Intracellular analysis of IFN-γ secretion. Around day 23 of the OP9/DL1 co-culture, generated cells from iPS cells were collected by vigorous pipetting through a 70-μm nylon mesh. Harvested cells were stimulated as described in methods, and Golgi stop solution was added to the culture 6 h before the analysis. **f** Survival curve of DLI mice model. First, mice got BMT. Fourteen days later, lymphoma cells inoculated into mice. On the following day, iPS-T cells were administered into mice. For double treated mice, another iPS-T cell treatment was administered seven days later from first treatment. By iPS-T cell treatment, mice survival was significantly prolonged compared with mice that recieved BMT only (BMT only vs. iPS-T cell single treatment: $p < 0.05$, BMT only vs. iPS-T cell double treatment: $p < 0.01$). Treatment effect was dependent on the number of treatment. **g, h** Mixed lymphocyte reaction analysis. Alloantigen reactivity of T cells was analyzed. iPS-T cells or splenic T cells from mice which have same background as iPS-T cells were stimulated with antigen-presenting cells from B6 mice or C3H mice. T cell proliferation was estimated by [^{3}H] thymidine uptake. Each T cell proliferation was shown in stimulation index. Stimulation index was calculated by dividing [^{3}H] thymidine uptake by T cells stimulated with B6 mice. **i, j, k** Frequencies of immune cell were evaluated after DLI treatment. CD4$^+$ T cells, CD8$^+$ T cells, and NK1.1$^+$ cells in spleen in lethally irradiated recipient mice which got BMT or BMT plus iPS-T cell administration were analyzed. In spleen, CD4$^+$ T cells were at comparable numbers in each group. In contrast, CD8$^+$ cells were significantly increased ($p < 0.05$), and NK1.1$^+$ cells were not significantly increased but tended to increase ($p = 0.056$) by administration of iPS-T cells

6.1.1 Repeated DLI Treatment Prolongs Mouse Survival

To test this new DLI treatment, we established a mouse model for lymphoma relapse after bone marrow transplantation (BMT), and then, we applied DLI treatment for this model. B6C3F1 mice (haplotype; b/k) were lethally irradiated, and then, bone marrow cells isolated from BDF1 mice (haplotype; b/d) were transplanted by intravenous injection. To mimic a lymphoma relapse, 6C3HED/OG mouse lymphoma cells [11] (haplotype; k) were inoculated intravenously two weeks after transplantation. On the following day, iPS-T cells were administered intravenously as a DLI treatment. In this experiment, mice that were given a single BMT died by day 28 after BMT. In contrast, mice survival was significantly prolonged when the mice got BMT plus DLI treatment, and treatment effect was depending on the treatment frequency (Fig. 6.1f).

Next we asked how this DLI therapy improved mice survival. One conceivable mechanism is allogeneic immune reaction by administered iPS-T cells against lymphoma cells. To examine this, we performed mixed lymphocyte reaction. iPS-T cells (H-2$^{b/d}$) were stimulated with CD11c$^+$ antigen-presenting cells from spleen of C57BL/6 (H-2^b) or C3H (H-2^k) mice. We found that the proliferation of iPS-T cells was not induced by CD11c$^+$ cells derived from C3H mice or C57BL/6 mice

(Fig. 6.1g, h). These results suggested that iPS-T cells generated in vitro did not have alloantigen reactivity. In our observation, CD4$^+$CD8$^+$ immature T cells in vivo did not react with allogeneic antigen-presenting cells (data not shown). Indeed, major population of iPS-T cells induced in vitro was shown to be CD4$^+$CD8$^+$ immature T cells [5]. Therefore, iPS cells derived T-lineage cells might not show alloantigen reactivity. Next, we counted numbers of CD4$^+$ or CD8$^+$ T cells and NK cells in spleens harvested from iPS-T cells-administered mice. Frequencies of CD4$^+$ T cells were comparable between mice that received a single BMT or BMT plus iPS-T cells. However, the combined treatment with BMT and iPS-T cells was accompanied with a slight increase in NK1.1$^+$ cells and importantly a significant increase in CD8$^+$ T cell frequencies compared to the single BMT treatment (Fig. 6.1i, j, k). These results suggested that iPS-T cells administered in combination with BMT increase recipient CD8$^+$ T cell number in spleen, and it may be involved in lymphoma cell elimination from recipient body and thus prolonged mice survival. However, further analysis is needed to understand the related mechanisms.

Recent studies have also proposed another new method of immunotherapy using cell reprogramming technology. Antigen-specific T cell is a potent effector immune cell against infected cells and tumor cells. Cytokine stimulation is one of the traditional manner to obtain enough numbers of antigen-specific T cells. However, such method sometimes elicit T cell's exhaustion. Cell reprogramming technology may also help to overcome these problems. T cell receptor gene segments rearrange during T cell development in the thymus and have critical role in the recognition of a specific antigen. Cell reprogramming process alters epigenetic gene modification of the cells, but does not change gene sequence itself. Therefore, it anticipates that when iPS cells are generated from a certain antigen-specific T cells (T-iPS cells), redifferentiated T cells from T-iPS cells will express and have the same antigen specificity as originated T cells, and epigenetic modification status is supposed to be "fresh." This may help to overcome exhaustion because it has been suggested that one of the exhaustion mechanisms is epigenetic modification of effecter molecule gene [12]. Based on these concepts, several groups have tested the reprogramming of tumor antigen- and virus antigen- specific human CD8$^+$ T cells and thereafter redifferentiating into T cells [13, 14]. As a result, T cells were successfully obtained, and these T cells have same antigen specificity as their originated T cells, from T-iPS cells. Furthermore, the antigen specificity and reactivity of these T cells were guaranteed as confirmed by the response of generated T cells to the same antigen.

In conclusion, new immunotherapy using cell reprogramming technology is already developing, and the combination of current treatment methods with cell reprogramming technologies will open up a lot of possibilities, which will be applicable for many kinds of immunotherapy.

6.2 Materials

6.2.1 Cells and Cell Lines

1. **iPS cells**: 38C2 cells were kindly gifted from Prof. Shinya Yamanaka (Kyoto University, Japan).
2. **MEFs**: Preparation method for MEFs was described previously [1].
3. **OP9/DL1 cells**: This cell lines were purchased from RIKEN BioResource Center (Tsukuba, Japan).
4. **6C3HED/OG**: This lymphoma cell line is distributed by Cell Resource Center for Biomedical Research, Tohoku University (Sendai, Japan).

6.2.2 Reagent and Culture Ware for General Use

1. Phosphate-buffered saline (PBS), without Ca^{2+} and Mg^{2+}.
2. 0.25 % trypsin (Life technologies).
3. 70-μm cell strainer (BD Falcon).
4. Cytokines: Flt3 ligand, mrIL-7, and mrIL-2 (all form R&D systems).

   Cytofix/Cytoperm^{TM} Permeabilization kit (BD).

6.2.3 Culture Medium and Cell Culture

1. **iPS cells**: iPS cells were maintained in DMEM medium supplemented with 15 % FCS, 2 mM L-glutamine, 0.1 mM nonessential amino acids, 0.1 mM 2-mercaptoethanol, 10 U/ml of penicillin, and 100 μg/ml of streptomycin (all from Life Technologies) containing 100× of recombinant human LIF supernatant (Wako) on feeder layers of irradiated MEF in culture dishes.
2. **MEFs**: MEFs were generated from embryos on day 14 as described previously [1] and maintained in DMEM medium supplemented with 10 % FCS, 2 mM L-glutamine, 0.1 mM nonessential amino acids, 10 U/ml of penicillin, and 100 μg/ml of streptomycin.
3. **OP9/DL1 cells**: OP9/DL1 cells were maintained in α-MEM supplemented with 20 % FCS, 0.1 mM 2-mercaptoethanol, 0.1 mM nonessential amino acids, 1 mM sodium pyruvate, 10 U/ml penicillin, 100 μg/ml of streptomycin, and 2.2 g/l of sodium bicarbonate.
4. **iPS cells differentiation media**: Differentiation media were based on OP9/DL1 media. Some appropriate cytokines were added depending on differentiation stage.

5. **6C3HED/OG cells**: This cell line was maintained in RPMI-1640 medium supplemented with 10 % FCS, 2 mM L-glutamine, 0.1 mM nonessential amino acids, 0.1 mM 2-mercaptoethanol, 10 U/ml of penicillin, and 100 μg/ml of streptomycin.

6.2.4 Flow Cytometry

1. Antibodies: anti-mouse IFN-γ (Clone: XMG1.2) (BD Pharmingen).
2. Flow cytometry buffer: PBS (w/o Ca^{2+} and Mg^{2+}) with 0.5 % FCS and 2 mM of EDTA.
3. Flow cytometer: FACSCalibur (BD) or FC500 (Beckman Coulter).
4. Flow cytometry analysis software: Cell Quest Pro (BD) or Flowjo (Tomy Digital Biology).

6.3 Methods

6.3.1 Differentiation of T-Lineage Cells from Murine Pluripotent Stem Cells

Differentiation of iPS cells was induced with a withdrawal of LIF from the culture in a non-treated plastic dish.

1. Remove iPS cell media from culture dish and wash by PBS (w/o Ca^{2+} and Mg^{2+}).
2. Add trypsin (1 ml/10 cm dish) and incubate at 37 °C for 3–5 min.
3. To dissociate cell aggregate, pipette cells gently but quickly using 1-ml pipette.
4. To abrogate the enzyme activity, add 9 ml of OP9/DL1 media and then centrifuge at 400g for 5 min.
5. Remove supernatant, loosen a cell pellet, and then add an appropriate volume of OP9/DL1 media.
6. Count iPS cells (See **Note 2**). Re-suspend 5×10^4 cells by OP9 media and plate on bacteriological untreated 10-cm dish, and place it in a 37 °C and 5 % CO_2 (See **Note 3**).
7. By day 5 of culture, embryoid body-like round-shaped spheres will be observed. Collect culture media including spheres into tube and then centrifuge at 400g for 5 min.
8. Remove supernatant and wash by PBS; then centrifuge at 400g for 5 min and remove supernatant.
9. To disrupt sphere, add 0.25 % trypsin and incubate at 37 °C for 5 min and then pipette cells gently but quickly using 1-ml pipette.
10. To abrogate the enzyme activity, add 9 ml of OP9 media and then centrifuge at 400g for 5 min.

11. Sample part of them and check expression of SSEA-1, Flk-1, and CD34 by flow cytometry (See **Note 4**).

12. Count cell number. Single-cell suspensions were replated at a density of 6×10^5 cells per 10 ml of culture media into 10-cm dish containing fresh semi-confluent OP9/DL1 cells with the addition of Flt3 ligand (5 ng/ml), and place it in a 37 °C and 5 % CO_2.

13. At day 8 of culture, harvest loosely adherent hematopoietic cells by gentle pipetting.

14. Sample part of them, check expression of CD45 by flow cytometry (See **Note 5**).

15. Centrifuge at 400g for 5 min, remove supernatant, add fresh OP9/DL1 medium with Flt3 ligand and exogenous IL-7 (5 ng/ml each), and then transfer cell suspension onto culture dish containing fresh semi-confluent OP9/DL1 cells.

16. Every 6 days thereafter, collect non-adherent iPS cell-derived hematopoietic cells by vigorous pipetting, filtered through a 70-μm nylon mesh and transferred onto fresh OP9/DL1 monolayers in OP9/DL1 media. On day 8 of culture, add another Flt3 ligand and exogenous IL-7 (5 ng/ml each). Both cytokines are included at all subsequent passages.

17. Sample part of them, check expression of CD25 and CD44 by flow cytometry (See **Note 6**).

18. Around 20–28 days thereafter, if the culture succeeded, T-lineage cells can be obtained, including CD4 and CD8 double-positive T cells, and small number of CD8 single-positive T cells (iPS-T cells).

6.3.2   *Functional Analysis of IPS-T Cells*

For example, IFN-γ production can be evaluated as a representative of T cell function.

1. Around 20 days after starting the differentiation culture, collect iPS-T cells from OP9/DL1 co-culture by vigorous pipetting and filtered through a 70-μm nylon mesh.

2. Re-plated the collected cells into a fresh culture dish and cultured for 1 h at 37 °C in 5 % CO_2 incubator to avoid contamination of OP9/DL1 cells.

3. To stimulate iPS-T cells, culture iPS-T cells for 2 days with plate-bound anti-CD3 (1 μg/ml; clone 145-2C11, Biolegend) monoclonal antibody (mAb) in differentiation medium in the presence of IL-2 (1 ng/ml) and anti-CD28 (1 μg/ml; clone 37.51, Biolegend) at 37 °C in 5 % CO_2 incubator.

4. Thereafter, add BD GolgiStop (BD) into the culture.

5. Incubate for 6 h in the presence of phorbol myristate acetate/ionomycin.

6. Perform intracellular staining for IFN-γ using anti-IFN-γ antibody and Cytofix/Cytoperm® (BD) according to the manufacturer's instructions.

7. Analyze cells by flow cytometer.

6.3.3   Preparation of Bone Marrow Cells for Bone Marrow Transplantation in DLI Model

1. Collect bone marrow cells from BDF1 mice.
2. Remove $CD3^+$ cells from bone marrow cells using biotin-conjugated anti-CD3 antibody and anti-biotin microbeads (Miltenyi Biotec).
3. Use CD3 negative bone marrow cells for BMT.

6.3.4   Treatment of DLI Mouse Model Using T-Lineage Cells from IPS Cells

1. Lethally irradiate B6C3F1 mice (haplotype; b/k) (e.g., 10 Gy irradiation using 0.2 mm of Cu and 1 mm of Al filter).
2. Transplant 2×10^6 cells of CD3 negative bone marrow cells into BDF1 mice (haplotype; b/d) intraintravenously.
3. Two weeks after transplantation administer 1×10^4 of 6C3HED/OG cells as a lymphoma relapse.
4. The following day, administer 1×10^6 of iPS-T cells (haplotype; b/d) intravenously.

6.4   Notes

1. All instruments and solutions used should be sterile.
2. iPS cells and MEFs can discriminate each other by their size
3. Cell number is critical for differentiation. This depends on cell lines. Too many iPS cells leave iPS cells in an undifferentiated state.
4. More than ~ 30 % over proportion of Flk-1 positive cells and minimum proportion of SSEA-1 positive cells are ideal. When you got large proportion of SSEA-1 positive cells, cell number may too high at starting point. Flk-1 positive cells give rise to $CD34^+$ cells.
5. If the culture succeeded, $CD45^+$ cells can be observed in the culture.
6. The stage of T cell development can be monitored by analyzing CD25 and CD44 expression. For example, $CD44^+ CD25^-$: DN1 stage, $CD44^+ CD25^+$: DN2 stage, $CD44^- CD25^+$ DN3 stage, and $CD44^- CD25^-$: DN4 stage.

References

1. Takahashi K, Yamanaka S. Induction of pluripotent stem cells from mouse embryonic and adult fibroblast cultures by defined factors. Cell. 2006;126(4):663–76. doi:10.1016/j.cell.2006.07.024.
2. Mellman I, Coukos G, Dranoff G. Cancer immunotherapy comes of age. Nature. 2011;480 (7378):480–9. doi:10.1038/nature10673.
3. Drake CG. Combination immunotherapy approaches. Ann Oncol. 2012;23(Suppl 8):viii41-46. doi:10.1093/annonc/mds262.
4. Chang YJ, Huang XJ. Donor lymphocyte infusions for relapse after allogeneic transplantation: when, if and for whom? Blood Rev. 2013;27(1):55–62. doi:10.1016/j.blre.2012.11.002.
5. Wada H, Kojo S, Kusama C, Okamoto N, Sato Y, Ishizuka B, Seino K. Successful differentiation to T cells, but unsuccessful B-cell generation, from B-cell-derived induced pluripotent stem cells. Int Immunol. 2011;23(1):65–74. doi:10.1093/intimm/dxq458.
6. Yamaguchi TP, Dumont DJ, Conlon RA, Breitman ML, Rossant J. flk-1, an flt-related receptor tyrosine kinase is an early marker for endothelial cell precursors. Development. 1993;118 (2):489–98.
7. Kabrun N, Buhring HJ, Choi K, Ullrich A, Risau W, Keller G. Flk-1 expression defines a population of early embryonic hematopoietic precursors. Development. 1997;124(10):2039–48.
8. Nakano T, Kodama H, Honjo T. Generation of lymphohematopoietic cells from embryonic stem cells in culture. Science. 1994;265(5175):1098–101.
9. Schmitt TM, Zuniga-Pflucker JC. Induction of T cell development from hematopoietic progenitor cells by delta-like-1 in vitro. Immunity. 2002; 17(6):749–756. doi: S1074761302004740.
10. Koch U, Fiorini E, Benedito R, Besseyrias V, Schuster-Gossler K, Pierres M, Manley NR, Duarte A, Macdonald HR, Radtke F. Delta-like 4 is the essential, nonredundant ligand for Notch1 during thymic T cell lineage commitment. J Exp Med. 2008;205(11):2515–23. doi:10.1084/jem.20080829.
11. Sobin LH, Kidd JG. Alterations in protein and nucleic acid metabolism of lymphoma 6C3HED-og cells in mice given guinea pig serum. J Exp Med. 1966;123(1):55–74.
12. Janson PC, Marits P, Thorn M, Ohlsson R, Winqvist O. CpG methylation of the IFNG gene as a mechanism to induce immunosuppression [correction of immunosupression] in tumor-infiltrating lymphocytes. J Immunol. 2008;181(4):2878–2886. doi:10.4049/jimmunol.181.4.2878
13. Vizcardo R, Masuda K, Yamada D, Ikawa T, Shimizu K, Fujii S, Koseki H, Kawamoto H. Regeneration of human tumor antigen-specific T cells from iPSCs derived from mature CD8 (+) T cells. Cell Stem Cell. 2013;12(1):31–6. doi:10.1016/j.stem.2012.12.006.
14. Nishimura T, Kaneko S, Kawana-Tachikawa A, Tajima Y, Goto H, Zhu D, Nakayama-Hosoya K, Iriguchi S, Uemura Y, Shimizu T, Takayama N, Yamada D, Nishimura K, Ohtaka M, Watanabe N, Takahashi S, Iwamoto A, Koseki H, Nakanishi M, Eto K, Nakauchi H. Generation of rejuvenated antigen-specific T cells by reprogramming to pluripotency and redifferentiation. Cell Stem Cell. 2013;12(1):114–26. doi:10.1016/j.stem.2012.11.002.

Chapter 7
Reprogramming of Human Cord Blood CD34⁺ Cells into Induced MSCs

Amanda Neises, Ruijun Jeanna Su and Xiao-Bing Zhang

Abstract Blood cells are the most accessible cells for reprogramming. Mesenchymal stem cells (MSCs) have multiple applications in regenerative medicine. We reported that induced MSCs (iMSCs) can be efficiently generated with OCT4 under defined conditions and iMSCs can be immortalized by overexpression of OCT4. Here, we detail the protocol for generating integration-free iMSCs from cord blood CD34⁺ cells using an episomal vector and describe the approach to immortalizing iMSCs.

Keywords Hematopoietic stem/progenitor cells · CD34⁺ cells · Mesenchymal stem cells · Reprogramming · OCT4 · Episomal vector

7.1 Introduction

One of the most exciting technological breakthroughs in the last decade is reprogramming somatic cells into pluripotency by ectopic expression of Yamanaka factors [1]. This discovery makes it possible to generate unlimited patient-specific cells for replacement therapy. We and other investigators have developed episomal vectors to efficiently and economically generate integration-free iPSCs from blood cells, the most accessible cell source in the human body [2–6]. These cells may have a greater potential for differentiation into hematopoietic lineages due to epigenetic memory [7]. iPSCs can also be efficiently differentiated into MSCs to treat multiple diseases such as diabetes and skeletal diseases [8].

A. Neises · R.J. Su · X.-B. Zhang (✉)
Department of Medicine, Loma Linda University, Loma Linda, CA 92354, USA
e-mail: xzhang@llu.edu

X.-B. Zhang
Division of Regenerative Medicine MC1528B, Loma Linda University,
11234 Anderson Street, Loma Linda, CA 92354, USA

© The Author(s) 2015
T. Cheng (ed.), *Hematopoietic Differentiation of Human Pluripotent Stem Cells*,
SpringerBriefs in Stem Cells, DOI 10.1007/978-94-017-7312-6_7

An alternative to the generation of iPSCs followed by direct differentiation into somatic stem cells like MSCs is direct reprogramming, which sidesteps the iPSC generation. We found that OCT4, one of the Yamanaka factors, is sufficient to reprogram human cord blood hematopoietic progenitor CD34$^+$ cells into induced MSCs (iMSCs) in 1–2 weeks after lentiviral transduction [9]. We also found that the activation of WNT signaling with GSK3 inhibitor CHIR99021 substantially increases the expansion of iMSCs [9]. Furthermore, we can also generate integration-free iMSCs from CD34$^+$ cells using an OCT4-expressing episomal vector. This vector is a plasmid containing EBNA and oriP elements from the EBV virus, which allow for short-term replication of plasmids in mammalian cells and a small portion of the plasmid is lost after each cell division [10, 11]. In our system, we found that OCT4 expression is undetectable and the plasmid is largely depleted at 3 weeks after nucleofection. This clinically relevant reprogramming approach may eventually find its way into clinical applications.

Our success in generating integration-free iMSCs is largely due to the optimized design of episomal vectors, including the use of the SFFV promoter, which substantially increases transgene expression levels in hematopoietic cells [10, 11]. We also optimized the culture conditions for MSCs, such as the use of fibronectin, hypoxia, and growth factors that increase MSC proliferation [9]. All of these improvements have enabled us to expand integration-free iMSCs for 10–15 passages. However, even under these culture conditions, we failed to culture integration-free iMSCs long term. Further investigation into the reprogramming process led to the finding that high-level expression of OCT4 is critical for reprogramming, but it strongly inhibits MSC differentiation. However, low-level OCT4 expression at ~ 5 % of that expressed in iPSCs can effectively inhibit spontaneous differentiation of iMSCs, while iMSCs are still able to differentiate into progeny, like osteoblasts when cultured in differentiation induction conditions. To achieve constitutive low-level OCT4 expression in iMSCs, cells were transduced with a lentiviral vector Lenti Sca1-OCT4, in which OCT4 expression is driven by a weak promoter, mouse Sca1. The transduced cells, termed Sca-O iMSCs, can be cultured for more than 30 passages and can be differentiated into adipocytes, osteoblasts, and chondrocytes. These immortalized cells express typical MSC markers such as CD73, CD90, and CD105 without showing any evidence of transformation. As follows we will detail the protocol for generating integration-free iMSCs and Sca-O iMSCs from cord blood.

7.2 Materials

7.2.1 Cord Blood CD34$^+$ Cells

1. Human cord blood. Drawn from umbilical cord and placenta after delivery and after obtaining informed consent.
2. CD34$^+$ cells MACS purification kit (Miltenyi; Cat. No. VPA-1003).

3. Red blood cell lysis buffer (RBC lysis buffer): weigh out 4.14 g of NH$_4$Cl and 0.5 g of KHCO$_3$. Dissolve the NH$_4$Cl and KHCO$_3$ in 450 ml of water and add 0.9 ml of 5 % EDTA. Adjust the pH to 7.2–7.4. Add water to bring the volume up to 500 ml and filter sterilize through a 0.2-μm filter. Store at 4 °C, but bring to room temperature before use.

4. Alternatively, cord blood CD34$^+$ cells can be purchased from Allcells (Cat. No. CB005F).

7.2.2 Cell Culture

1. HSC culture medium: SFEM (Stemcell Technologies; Cat. No. 09650) was supplemented with 100 ng/ml SCF (Prospec, East Brunswick, NJ; Cat. No. CYT-255), 100 ng/ml TPO (Prospec; Cat. No. CYT-302), 100 ng/ml Flt3 Ligand (Prospec; Cat. No. CYT-331), 10 ng/ml IL-3 (Prospec; Cat. No. CYT-210), 10 ng/ml G-CSF (Prospec; Cat. No. CYT-220), 0.1 mM 2-Mercaptoethanol (ME) (Sigma-Aldrich; Cat. No. M6250), 1 % penicillin/streptomycin (P/S) (ABM; Cat. No. G255), 1 % ITS (Life Technologies; Cat. No. 41400-045), and 200 μm ascorbic acid 2-phosphate (Sigma-Aldrich; Cat. No. 49752). Store the medium at 4 °C for up to 3 weeks.

2. iMSC generation and expansion medium: MEM α (Life Technologies; Cat. No. 32561-037) medium was supplemented with 5 % Knockout serum replacement (Invitrogen; Cat. No. 10828-028), 2 % fetal bovine serum (FBS) (ABM; Cat. No. TM999-500), 1 % P/S, 20 ng/ml FGF2 (ABM; Cat. No. Z101455), 20 ng/ml EGF (ABM; Cat. No. Z200025), 20 ng/ml PDGF-BB (ABM; Cat. No. Z100355) 1 % ITS, and 200 μm ascorbic acid 2-phosphate. Store at 4 °C for up to 3 weeks.

3. Accutase (Innovative Cell Technologies; Cat. No AT-104) (see **note 1**).

4. Human fibronectin (BD Biosciences; Cat. No. 354008). Add 50 ml of PBS supplemented with 1 % P/S to 1 mg of fibronectin. Sterilize by filtering through a 0.22-μm syringe filter. Store at 4 °C for up to 2 months. May be aliquoted and frozen down at −20 °C for long-term storage. Before generating or passaging iMSCs, add 1 ml of fibronectin medium into each well of the non-tissue culture (TC)-treated 6-well plates. Incubate at 37 °C for 2 h. Aspirate out fibronectin medium and add medium and cells for culture (see **note 2**).

5. CHIR99021. Small molecule inhibitor of GSK3, CHIR99021, was purchased from ABM (Cat. No. G611). The final concentration used for cell culture is 3 μM (see **note 3**).

6. 2× Freezing medium. Dissolve 5 g of trehalose (Sigma-Aldrich Cat. No. T9531) in 30 ml of water in a 37 °C water bath, bring the temperature to 4 °C, and then add 10 ml of FBS and 10 ml of DMSO (Fisher Scientific; Cat. No. BP231-1) [12]. Filter sterilize with a 0.22-μm syringe filter. Store at 4 °C for up to 3 months. Add equal volume of freezing medium to the cell suspension to freeze down the cells for storage.

7.2.3 Episomal Vector and Nucleofection Kits

1. Episomal vector: pEV SFFV-OCT4-wpre (pEV-O) (see **note 4**).
2. Nucleofector™ Kits for Human CD34⁺ Cells (Lonza, Walkersville, MD; Cat. No. VPA-1003).

7.2.4 Lentiviral Vector and Transduction

1. Lentiviral vector: Lenti-Sca1-OCT4-wpre (Sca-O).
2. Lentiviral vector packaging: Follow standard protocol that is detailed elsewhere [13].
3. Protamine sulfate (Sigma-Aldrich; Cat. No. P3369). Make a stock of 4 mg/ml by dissolving 400 mg protamine sulfate in 100 ml PBS. Filter sterilize with a 0.22-μm syringe filter and aliquot and keep in −20 °C for long-term storage. The working concentration is 8 μg/ml for lentiviral transduction (see **note 5**).

7.2.5 Plastics and Equipment

1. Non-TC-treated 6-well plates (BD Falcon; Cat. No 351146).
2. 5- or 15- or 50-ml polystyrene tubes (BD Falcon).
3. Pipettes and pipettors.
4. Nucleofector (Lonza; Amaxa II).
5. CO_2 incubator.
6. Centrifuge.
7. Hypoxia chamber (Stemcell Technologies; Cat. No. 27310).
8. Mixed gas cylinder. Order a mixed gas composed of 3 % O_2, 5 % CO_2, and 92 % N_2. Flush the hypoxia chamber at 30 L/min for 1 min (see **note 6**).
9. Flow cytometer.
10. Inverted microscope.

7.2.6 Immunohistochemistry and Flow Cytometry

1. Flow cytometry (FACS) buffer: PBS supplemented with 2 % FBS. Store at 4 °C. For long-term storage, add 0.05 % sodium azide (Sigma-Aldrich; Cat. No. S8032).
2. Fixation buffer and permeabilization buffer (eBioscience, San Diego, CA; Cat. No. 00-8222-49 and 00-8333-56).

3. Antibodies: Nestin-Alexa Fluor® 647 (BD Pharmingen; Cat. No. 560341). All the other antibodies were purchased from eBioscience (San Diego, CA). CD14-FITC (11-0149-42), CD31 (PECAM-1)-FITC (11-0319-42), CD34-PE (12-0349-42), CD45 FITC (11-9459-42), CD73-PE (12-0739-42), CD105-PE (12-1057-42), OCT4-PE (12-5841-82), TRA-1-60-PE (12-8863-82).

7.2.7 Trilineage Differentiation of iMSCs

1. Adipocytic differentiation medium. MEM α (Life Technologies; Cat. No. 32561-037) was supplemented with 1 μm dexamethasone (Sigma-Aldrich; Cat. No. D4902), 1 μm Troglitazone (Sigma-Aldrich; Cat. No. T2573), 10 μg/ml insulin (Sigma-Aldrich; Cat. No. 91077C), 0.5 mM isobutylxanthine (Sigma-Aldrich; Cat. No. I5879), 5 ng/ml of FGF2 (ABM), 10 % FBS, and 1 % P/S.
2. Osteoblastic differentiation medium. MEM α was supplemented with 0.1 μm dexamethasone, 200 μm ascorbic acid 2-phosphate, 10 mM β-Glycerophosphate (Sigma-Aldrich; Cat. No. G9422), 10 ng/ml BMP2, 10 ng/ml BMP4, 10 % FBS, and 1 % P/S.
3. Chondrogenic differentiation medium. MEM α was supplemented with 0.1 μm dexamethasone, 200 μm ascorbic acid 2-phosphate, 5.33 μg/ml linoleic acid (Sigma-Aldrich, Cat. No. L1012), 0.35 mm L-proline (Sigma-Aldrich, Cat. No. P5607), 10 ng/ml TGFβ3 (Stemgent, San Diego, CA), 10 ng/ml TGFβ1 (Stemgent) and 1 % ITS, 10 % FBS, and 1 % P/S.
4. Oil Red O staining. Adipocytes were stained with Oil Red O solution. The stock solution was made by dissolving 0.5 g of Oil Red O (Sigma-Aldrich; Cat. No. O0625) powder in 100 ml isopropanol (Sigma-Aldrich; Cat. No. I9516) with gentle heat and then filtered with a 5-μm syringe filter. Oil Red O working solution was made by diluting Oil Red O stock solution with nanopure water at a ratio of 3:2.
5. Alizarin Red staining. Bone nodule formation was evaluated by Alizarin Red staining. The staining solution was made by dissolving 2 g of Alizarin Red S (Sigma-Aldrich; Cat. No. A5533) in 100 ml of nanopure water. The pH was adjusted to 4.3 with 10 % ammonium hydroxide (Sigma-Aldrich; Cat. No. 320145) and then filtered with a 5-μm syringe filter.
6. Alcian Blue staining solution was prepared by dissolving 1 g of Alcian Blue (Sigma-Aldrich; Cat. No. A5268) in 100 ml of 3 % acetic acid (Sigma-Aldrich, Cat. No. 320099) solution. The pH was adjusted to 2.5 using acetic acid.

7.3   Methods

Conduct all experiments at room temperature unless otherwise specified.

7.3.1   CD34 Enrichment

1. Add 4–6× volume of RBC lysis buffer to the cord blood, and the total volume should be less than 30 ml in each 50-ml conical tube (see **note 7**). If there is more than 30 ml, split into 2 tubes. Incubate at room temperature for 5 min.
2. Centrifuge the cells at 400 g for 10 min and remove the supernatant. Resuspend the cell pellet in 20 ml of IMDM, count the number of cells, and then centrifuge the cells again at 400 g for 5 min.
3. Resuspend the cell pellet in up to 1×10^9 cells/ml with Miltenyi MACS Separation Buffer, add CD34 microbeads, and enrich CD34$^+$ cells per manufacturer's instructions.
4. Resuspend the cells in 2 ml of HSC medium and count the number of cells.

7.3.2   CD34 Cell Culture

1. Culture the cells in HSC medium at a density of 0.5×10^6 cells/ml. Add 2 ml to each well of a non-TC-treated 6-well plates.
2. Culture the cells at 37 °C for 2–4 days in a 5 % CO_2 incubator with a water tray to maintain the humidity.

7.3.3   Nucleofection

1. Add 2 μg pEV-O plasmid in a sterile Eppendorf tube, resuspend in 5 μl H_2O (see **note 8**).
2. Heat the tube at 50 °C for 5 min (see **note 9**). Cool down the tube to room temperature and add 74 μl of nucleofection buffer and 16 μl of the supplement provided by the Nucleofector™ Kits.
3. Mix the DNA and the buffer with a 100 μl tip and add to the cell pellet. Flick the bottom of the tube, using your finger, 3–5 times to resuspend the cells in the buffer (see **note 10**).
4. Use the kit-provided plastic pipette to transfer the mixture to the provided cuvette without producing any bubbles. Use the program U-008 for nucleofection.

5. Before nucleofection, prepare a fibronectin-precoated non-TC 6-well plates. Add 1.5- to 2-ml HSC medium into each well. Prewarm the culture plate in the 37 °C incubator.
6. After nucleofection, use the same pipette to transfer 500-μl medium from the culture well to the cuvette. Place the pipette to the bottom of the cuvette and transfer the cells to the culture well.
7. Place the plate in a hypoxia chamber, flush with the hypoxia gas for 1 min, and transfer the chamber into the incubator (see **note 11**).

7.3.4   Generation of Integration-Free iMSCs

1. On day 2, add 2 ml of iMSC generation medium into each well and add 3 μM CHIR99021 for 1 week.
2. Starting on day 4, change the MSC medium every 2 days (see **note 12**).
3. Small iMSC-like colonies will appear 3–5 days after nucleofection.
4. At one week after nucleofection, when the colonies cover ∼50 % of the plate surface, passage cells by adding 600 μl of Accutase to each well. Transfer cells to fibronectin-precoated non-TC 6-well plates. Split the cells at a ratio of 1:4–6 every 2–3 days.
5. After 2–4 passages, cells may be immortalized by low-level OCT4 overexpression.

7.3.5   Immortalization of iMSCs

1. Seed 5×10^5 iMSCs in 1-ml MSC medium into one well of a fibronectin-precoated non-TC 6-well plates. Add packaged Lenti Sca1-OCT4 at a multiplicity of infection (MOI) of 1. Add protamine sulfate at a final concentration of 8 ug/ml and place the plate back to the incubator.
2. Four to 8 h later, aspirate out the medium and add 2 ml of fresh MSC medium for continued culture. Once cells reach 80–100 % confluence, split the cells at a ratio of 1:4–6.

7.3.6   Long-Term Culture of iMSCs

1. For long-term passage of iMSCs, precoat the non-TC 6-well plates with fibronectin.

2. When the cells reach 80–100 % confluency, passage the cells by aspirating out the spent medium and adding 600 µl of Accutase to each well.
3. Incubate at 37 °C for 5 min. Transfer 100–150 µl of cells to a new well for continued culture under hypoxia.

7.3.7   Freezing Down iMSCs

1. After several passages, the iMSCs may be frozen down for long-term storage. At 2–3 days after the cell split, when iMSC reach a confluency of 80–100 %, aspirate out the spent medium and add 500 µl of Accutase. Incubate the cells at 37 °C for ∼5 min and then add 500 µl of 2× freezing medium before transferring the cells to a 1-ml cryovial.
2. Mix well and transfer the vial to a −80 °C freezer for short-term storage (days to weeks). Several hours later when the cells are frozen, the vials may be transferred to a liquid nitrogen tank for long-term storage (years).

7.3.8   Phenotyping of iMSCs by Flow Cytometry

1. For intracellular staining of OCT4 and Nestin, add 600 µl of Accutase to each well and incubate at 37 °C for 5 min. Pipette the cells up and down to make a single cell suspension. Use 100 µl for cell staining. Add 900 µl of fixation buffer and 100 µl of 10× permeabilization buffer and incubate the cells at room temperature for 10 min. After spinning down the cells at 400 g for 5 min, the cells were resuspended in 100 µl of FACS buffer with 10 % permeabilization buffer and stained with the antibody at room temperature for 2 h. Wash the cells twice with 2 ml of FACS buffer supplemented with 10 % permeabilization buffer and resuspend the cells in 200–300 µl of FACS buffer before conducting FACS analysis.
2. For staining cell surface markers like CD73 and CD105, cells resuspended in Accutase were incubated with antibodies for 30 min at room temperature. After washing the cells with 2 ml of FACS buffer and resuspending the cells in 200–300 µl of FACS buffer, flow cytometric analysis was performed using FACS Aria II (BD Biosciences, San Jose, CA). Thirty thousand events were collected for each sample (see **notes 13 and 14**).

7.3.9   Trilineage Differentiation of iMSCs

1. For the induction of adipocytic, osteoblastic, or chondrogenic differentiation, 1×10^5, 2×10^5, or 4×10^5 iMSCs were seeded in each well of TC-treated 6-well plates.

2. All cultures were maintained with 5 % CO_2 in a water-jacketed incubator at 37 °C, and culture media were changed every 2–3 days. Three to four weeks after differentiation culture, the cells were fixed in 10 % neutral-buffered formalin for 15 min before staining.
3. Adipocytes were stained with 1 ml of Oil Red O working solution for 15 min, and the stain was then washed out with 60 % isopropanol.
4. For bone nodule staining after osteoblastic differentiation, 1 ml of Alizarin Red staining solution was added for 5 min after formalin fixation, and the cells were washed with nanopure water.
5. To stain mucopolysaccharides associated with chondrocytic differentiation, fixed cultures were stained with Alcian Blue staining solution for 30 min. The cultures were later washed in tap water for 2 min and then rinsed in nanopure water (see **notes 15**).

7.4   Notes

1. Accutase is used for cell splitting. Alternatively, trypsin can be used to passage iMSCs. The use of Accutase simplifies the cell passage procedure, because one does not need to remove or neutralize the enzyme.
2. Fibronectin is very critical for successful reprogramming of blood cells into iMSCs. It can also increase the proliferation rate of iMSCs. Fibronectin may be reused multiple times without obvious decrease in its activity.
3. CHIR99021 synergizes with OCT4 to expand iMSCs. However, in the absence of OCT4, CHIR99021 may have a negative effect on MSC proliferation. CHIR99021 is encouraged to be used during the first week of iMSC reprogramming.
4. Use endo-free Maxiprep Kits to obtain high-quality plasmid DNA (Qiagen; Cat No.12362).
5. Protamine sulfate generally increases transduction efficiency by 20–30 %.
6. iMSCs may be cultured in a 5 % O_2 hypoxia incubator, which is controlled by flushing with N_2 and CO_2. We prefer the use of a hypoxia chamber flushed with 3 % O_2 and 5 % CO_2.
7. Keeping the RBC lysis buffer at 4 °C is encouraged for long-term storage. Bring it to room temperature before use. Long-term storage at room temperature may lead to deterioration or degradation of the buffer. The use of RBC lysis buffer instead of Ficoll gradient centrifugation may increase the yield of $CD34^+$ cells, while it does not obviously affect the quality of enriched $CD34^+$ cells.
8. The use of more than 5 μg DNA may significantly decrease cell survival after nucleofection.
9. Heat treatment of the plasmid can prevent *Escherichia coli* contamination in cell culture.

10. Our experience suggests that the optimal cell density for nucleofection is $1-2 \times 10^6$ cells. Too low or too high a density may decrease the reprogramming efficiency and/or affect cell survival.
11. A common phenomenon of cells cultured in 6-well plates is that more cells are located in the center of wells. To prevent uneven spreading, gently rock the plate front and back twice to mix the cells after adding medium and cells to the wells. Rock the plates again right before placing them back into the incubator.
12. When changing the culture medium, leave ~ 500 µl per well to prevent the drying out of these cells in the wells. This can also increase the efficiency, in particular, for large-scale cell culture.
13. Accutase does not obviously affect antigen–antibody interaction, thus antibodies can be added directly to Accutase-treated cells for staining.
14. iMSCs express typical MSC markers CD29, CD44, CD73, CD90, CD105, and CD166 and do not express hematopoietic markers CD14, CD34, and CD45, endothelial marker CD31, and iPSC markers OCT4 and TRA-1-60.
15. Compared to bone marrow-derived MSCs, cord blood-derived iMSCs appear to have a greater capacity for chondrocytic differentiation, but have an inferior ability for adipocytic differentiation.

References

1. Takahashi K, Yamanaka S. Induction of pluripotent stem cells from mouse embryonic and adult fibroblast cultures by defined factors. Cell. 2006;126(4):663–76.
2. Chou BK, Mali P, Huang X, et al. Efficient human iPS cell derivation by a non-integrating plasmid from blood cells with unique epigenetic and gene expression signatures. Cell Res. 2011;21(3):518–29.
3. Su R-J, Baylink DJ, Neises A, et al. Efficient generation of integration-free iPS cells from human adult peripheral blood using BCL-XL together with Yamanaka factors. PLoS ONE. 2013;8(5):e64496.
4. Liu SP, Li YX, Xu J, et al. An improved method for generating integration-free human induced pluripotent stem cells. Zhongguo Shi Yan Xue Ye Xue Za Zhi. 2014;22(3):580–7.
5. Su RJ, Neises A, Zhang XB. Generation of iPS cells from human peripheral blood mononuclear cells using episomal vectors. Methods Mol Biol. 2014.
6. Chou BK, Gu H, Gao Y, et al. A facile method to establish human induced pluripotent stem cells from adult blood cells under feeder-free and xeno-free culture conditions: a clinically compliant approach. Stem Cells Transl Med. 2015;4(4):320–32.
7. Kim K, Doi A, Wen B, et al. Epigenetic memory in induced pluripotent stem cells. Nature. 2010;467(7313):285–90.
8. Jung Y, Bauer G, Nolta JA. Concise review: induced pluripotent stem cell-derived mesenchymal stem cells: progress toward safe clinical products. Stem Cells. 2012;30(1):42–7.
9. Meng X, Su RJ, Baylink DJ, et al. Rapid and efficient reprogramming of human fetal and adult blood CD34(+) cells into mesenchymal stem cells with a single factor. Cell Res. 2013;23 (5):658–72.
10. Meng X, Neises A, Su R-J, et al. Efficient reprogramming of human cord blood CD34$^+$ cells into induced pluripotent stem cells with OCT4 and SOX2 alone. Mol Ther. 2012;20 (2):408–16.

11. Zhang XB. Cellular reprogramming of human peripheral blood cells. Genomics Proteomics Bioinf. 2013;11(5):264–74.
12. Zhang XB, Li K, Yau KH, et al. Trehalose ameliorates the cryopreservation of cord blood in a preclinical system and increases the recovery of CFUs, long-term culture-initiating cells, and nonobese diabetic-SCID repopulating cells. Transfusion. 2003;43(2):265–72.
13. Delenda C. Lentiviral vectors: optimization of packaging, transduction and gene expression. J Gene Med. 2004;6(Suppl 1):S125–38.

Chapter 8
CRISPR/Cas9-Mediated Genome Editing in Human Pluripotent Stem Cells

Jian-Ping Zhang, Amanda Neises, Tao Cheng and Xiao-Bing Zhang

Abstract Induced pluripotent stem cells (iPSCs) hold great promise for gene and cell therapies. Correction of a diseased gene is often required before conducting directed differentiation of iPSCs. In addition, creation of iPSC reporter lines greatly facilitates high-throughput screening and other applications. Recent advances in the CRISPR genome editing technology have made it possible to readily accomplish these goals. Here, we describe a step-by-step procedure to efficiently generate a GFP reporter iPSC line by using a Cas9-sgRNA vector and an optimized donor template plasmid.

Keywords Human pluripotent stem cells · CRISPR · Cas9-sgRNA · Genome editing · Knock-in

8.1 Introduction

The generation of human induced pluripotent stem cells (iPSCs) from adult somatic cells holds great potential for gene and cell therapies. iPSCs are considered an ideal source of autologous cells for cell replacement therapy because iPSCs can be induced to differentiate into more than 200 types of cells. iPSCs can be generated from fibroblasts derived from a skin biopsy or trace amount of renal proximal

J.-P. Zhang · T. Cheng · X.-B. Zhang (✉)
State Key Laboratory of Experimental Hematology, Institute of Hematology and Blood
Disease Hospital, Chinese Academy of Medical Sciences and Peking Union Medical College,
Tianjin 300020, China
e-mail: xzhang@llu.edu

A. Neises · X.-B. Zhang
Department of Medicine, Loma Linda University, Loma Linda, CA 92354, USA

X.-B. Zhang
Division of Regenerative Medicine MC1528B, Department of Medicine, Loma Linda
University, 11234 Anderson Street, Loma Linda, CA 92350, USA

© The Author(s) 2015   103
T. Cheng (ed.), *Hematopoietic Differentiation of Human Pluripotent Stem Cells*,
SpringerBriefs in Stem Cells, DOI 10.1007/978-94-017-7312-6_8

tubular cells present in urine [1]. We and other investigators have generated integration-free iPSCs using mononuclear cells from peripheral blood (PB) [2–6]. PB is the easy to access, minimally invasive, and the most abundant cell source in human body for generating iPSCs. PB-derived iPSCs may be particularly useful in generating blood products, as one can take advantage of the epigenetic memory of these cells [7].

For iPSC-based cell replacement therapy, one needs to correct the diseased gene before differentiating them into cells of clinical interest. On the other hand, to facilitate the efforts in identifying novel approaches for directed differentiation of iPSCs, one may need to generate an iPSC reporter cell line. A recent technological breakthrough has made it possible to edit the genome rapidly and economically. This exciting technology is called CRISPR, or the clustered regularly interspaced short palindromic repeat, which uses a single-guide RNA (sgRNA) in complex with a CRISPR-associated nuclease Cas9 to search and bind with the complementary sequence on the genome, followed by excision of double-stranded DNA at the precise target locus [8, 9].

The CRISPR-Cas9 system has been widely used in genome editing due to the ease of vector construction and high targeting efficiency. A single-guide RNA (sgRNA) targets Cas9 to genomic regions that are complementary to the 20-nucleotide (nt) target region of the sgRNA that contains a 5′-NGG-3′ protospacer-adjacent motif (PAM). The Cas9 nuclease generates a DNA double-strand break (DSB) three base pairs upstream of the PAM [10–12]. The DSB is often repaired by the error-prone non-homologous end-joining (NHEJ) pathway, which induces nucleotide substitution, insertions and deletions (indels) and may lead to frameshift mutations [13, 14]. Alternatively, the DSB may be repaired by high-fidelity homologous recombination (HR) if an exogenous DNA template flanked by homologous sequences to the target site is provided. Targeted editing allows for correction of disease-causing mutations by replacing a mutated gene directly or the insertion of a functional copy of the affected gene into a safe genomic harbor or downstream of its own promoter [15–18].

In an early report, the gene knock-in efficiency using the CRISPR-Cas9 system in human pluripotent stem cells was relatively low, ranging from 1 to 4 % [10]. To enhance its efficiency, investigators have optimized this approach by (a) optimization of transfection; (b) enrichment of the transfected cells; (c) optimizing the design of donor DNA template by flanking it with sgRNA target sites, thereby inducing the cleavage of a donor template plasmid by a Cas9–sgRNA complex; and (d) using NHEJ inhibitors to improve HR efficiency [19–23]. The specificity of CRISPR-Cas9 system was considered a major safety concern for clinical applications [24]; however, whole-genome sequencing studies showed low incidence of off-target mutations induced by Cas9-sgRNA in human iPSCs [25, 26]. Recently, we and other investigators have shown that targeting the specificity can be substantially increased by the use of truncated guide RNAs. When the length of sgRNA is reduced from 20 to 17 bp or 18 bp, the off-target efficiency can be reduced by 1000-fold [12].

Fig. 8.1 An outline of procedures to generate a GFP reporter human iPSC line at OCT4 locus

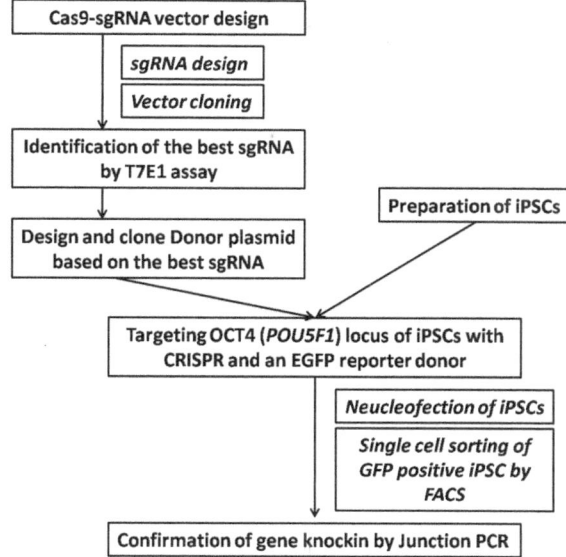

In this chapter, to exemplify genome editing in iPSC cells, we describe a step-by-step procedure to generate a GFP reporter line at the OCT4 (POU5F1) locus (Fig. 8.1). We recommend designing three sgRNAs to target the DNA surrounding the stop codon of OCT4, followed by identifying the best sgRNA with the highest cleavage efficiency. The cleavage efficiency will be determined by a T7E1 mutation detection assay, an enzyme mismatch cleavage method, after transfecting 293T cells with individual sgRNA together with Cas9. To target the C terminus of OCT4, the stop codon of OCT4 is deleted and the last OCT4 codon is fused in frame with a 2A-GFP-Wpre sequence, where 2A is a self-cleaving peptide. To enhance the knock-in efficiency, the sgRNA target sequence is also included to flank the donor template. After the GFP knock-in, GFP is under the control of the endogenous OCT4 promoter; thus, the GFP intensity reflects the OCT4 expression in cells. As follows, we describe in detail the optimized design of sgRNA targeting system to improve efficiency and specificity, culture of iPSCs, and nucleofection of iPSCs with sgOCT4 and a GFP reporter donor and conclude with the selection of the GFP-positive clones by flow cytometry.

8.2   Materials

8.2.1   Cell Culture

1. 293T culture medium. DMEM (Dulbecco's modified Eagle medium; HyClone, Cat. No. SH30243.01B) was supplemented with 10 % fetal bovine serum (FBS;

GIBCO, Cat. No. 16000-044) and 1 % P/S (penicillin/streptomycin; ABM, Cat. No. G255). Store the medium at 4 °C for up to 4 weeks.

2. iPSC culture medium. PB-derived iPSCs were generated in our laboratory. iPSCs were maintained in E8 medium (Essential 8 Medium; GIBCO, Cat. No. A15169-01).
3. 0.25 % trypsin/EDTA (GIBCO, Cat. No. 25200-056).
4. Accutase (GIBCO, Cat. No. A11105-01).
5. 0.5 mm EDTA (cell disassociation buffer). Dilute 500 μl of 0.5M EDTA (GIBCO, Cat. No. 15575) in 500 ml PBS.
6. ROCK inhibitor (Y-27632; ABM, Cat. No. G604).
7. Matrigel-coated plates. Dissolve 1 ml Matrigel (BD, Cat. No. 354230) in 50 ml cold Knockout DMEM/F12 medium (GIBCO, Cat. No. 12660-012). Add 1 ml diluted Matrigel to each well of 6-well plates. Incubate at 37 °C for 1 h or at 4 °C for overnight. Aspirate out the medium before cell culture.
8. T7E1 (Peking polymath technology, Cat. No. E001L).

8.2.2 Transfection Reagents and Nucleofection Kits

1. Lipofectamine 2000 (Invitrogen, Cat. No. 11668-019).
2. Human Stem Cell Nucleofector® Kit (Lonza, Cat. No. VPH5022).

8.2.3 DNA Extraction, PCR, and Cloning

1. TaKaRa MiniBEST Universal Genomic DNA Extraction Kit (Takara, Cat. No. 9765).
2. QIAquick PCR purification kit (Qiagen, Cat. No. 28104).
3. 2× FastPfu PCR SuperMix (TransGen, Cat. No. as221-1).
4. KAPA HiFi polymerase (Kapa Biosystems, Cat. No. KK2102).
5. Phusion® High-Fidelity PCR Master Mix with HF Buffer (NEB, Cat. No. M0531).
6. CloneJET PCR Cloning Kit (Thermo, Cat. No. K1232).
7. Gibson Assembly® Cloning Kit (NEB, Cat. No. E5510S).

8.2.4 Plastics and Equipment

1. TC-treated 6-well plates (BD Falcon; Cat. No. 353046).
2. 5- or 15- or 50-ml polystyrene tubes (BD Falcon).

3. Pipettes and pipettors.
4. Nucleofector (Lonza; Amaxa II).
5. CO_2 incubator.
6. Centrifuge.
7. PCR thermal cycler.
8. FACSAria II flow cytometer.
9. Inverted microscope.

8.3 Methods

8.3.1 sgRNA Design and Vector Cloning

8.3.1.1 sgRNA Design

1. Retrieve the sequence of ~100 bp surrounding OCT4 stop codon by using Genome Browser (http://genome.ucsc.edu/cgi-bin/hgBlat?command=start).
2. Design regular 20-bp sgRNAs at CHOPCHOP Web site (https://chopchop.rc. fas.harvard.edu/) by pasting the input sequence. Truncate the length of the sgRNA to 17 or 18 bp by deleting 2–3 bases at the 5′ end (see note 1).
3. Preferentially pick sgRNAs with a G or an A at the 3′ end. The targeting sequences are (N16)GNGG, (N16)ANGG, (N16)CNCC, or (N16)TNCC (see note 2).
4. Preferentially pick sgRNAs with a G at the 5′ end. If the first nucleotide of sgRNA is an A, C, or T, add a G in front of it, as U6 promoter-mediated transcription starts at a G.
5. Search off-target sites at TagScan (http://ccg.vital-it.ch/tagger/tagscan.html). For 17-bp sgRNAs, abandon the ones that have a perfect match at other loci of the human genome. For 18-bp sgRNAs, abandon the ones that have less than 1 mismatch at other loci of the human genome.

8.3.1.2 Cas9-sgRNA Vector Cloning

6. Use plasmid pU6-sgBbsI-EF1-Cas9-E2A-Puro-wpre to provide the sgRNA vector backbone.
7. Digest pU6-sgBbsI-EF1-Cas9-E2A-Puro-wpre vector with BbsI enzyme at 37 ° C overnight (see note 3).

Prepare the following mixture:

Plasmid	5 μg
BbsI	3 μl
10× buffer	5 μl
H$_2$O	Bring total volume to 50 μl

Run the PCR product on a 1 % agarose gel at 70 V for 2 h. Cut out the correct band (9.5 kb) and purify the PCR products with the QIAquick PCR purification kit according to the manufacturer's instructions.

8. Synthesize sgRNA template: TATATATCTTGTGGAAAGGACGAAACAC CG NNNNNNNNNNNNNNNN GTTTTAGAGCTAGAAATAGCAAGTT AAAAT. PCR primers are listed as follows: sgRNA-F: TATATATCTTGTGG AAAGGACGAA and sgRNA-R: ATTTTAACTTGCTATTTCTAGCTCTAA. Use the KAPA HiFi polymerase to amplify the sgRNA product, using the following cycling conditions: 98 °C for 2 min, 1 cycle; 98 °C for 5 s, 60 °C for 20 s, 20 cycles. Purify the PCR products with the QIAquick PCR purification kit according to the manufacturer's instructions.
9. Assemble 100 ng of the sgRNA backbone and 10 ng of the sgRNA PCR product using Gibson Assembly® Cloning Kit in a total volume of 20 μl. Incubate in a thermocycler at 50 °C for 15 min. After incubation, use 1 μl for transformation. After overnight culture at 37 °C, pick 3 clones into 15-ml tubes with 1–2 ml LB medium and shake at 37 °C for 8–16 h.
10. Isolate the plasmid DNA from cultures by using a QIAprep Spin Miniprep Kit according to the manufacturer's instructions. Verify the sequence of each clone by Sanger sequencing. The sequencing primer is U6-F: GGGCAGGAAGAGGGCCTAT.
11. Pick the clone with correct insert. Culture 100 μl *Escherichia coli* in 250 ml LB medium for 16–18 h, followed by extracting plasmids with Qiagen Plasmid Plus Maxi Kit.

8.3.2 Identification of the Best sgRNA by T7E1 Assay

8.3.2.1 Preparation of 293T Cells for Transfection

1. 293T cells are maintained in 6-well plates. Passage cells when reaching 80–90 % confluency.
2. To passage, remove the medium and add 2 ml PBS to rinse the cells.
3. Remove the PBS, add 0.4 ml of 0.25 % trypsin/EDTA to each well, and incubate at 37 °C.

4. After 3–5 min of incubation, add 1 ml of warm DMEM/10 % FBS to inactivate the trypsin, dissociate the cells by pipetting them up and down gently, and then transfer the cells to a 15-ml tube.
5. Count the cells using a hemacytometer.
6. Seed the cells into 24-well plates at a density of $1–2 \times 10^5$ per well in a total volume of 500 μl (see note 4).

8.3.2.2   Transfection of 293T Cells Using Lipofectamine 2000

7. One day later, transfect 0.5 μg of the Cas9-sgOCT4 plasmid and include a GFP plasmid control to 293T cells, following Lipofectamine 2000 manufacturer's instructions.
8. Add lipofectamine and DNA mix to the cells gently.
9. Check the transfection efficiency after 24 h by using a fluorescence microscope (if the GFP control plasmid was cotransfected).
10. Incubate the cells for a total of 72 h after transfection.

8.3.2.3   Harvesting Cells for DNA Extraction

11. 72 h after transfection, dissociate the transfected cells and harvest them by centrifugation at 1000 rpm for 5 min at room temperature.
12. Aspirate out the medium completely, leaving the cell pellet at the bottom of the tube.
13. Isolate the genomic DNA using Genomic DNA Extraction Kit (Qiagen) according to the manufacturer's instructions.

8.3.2.4   Amplifying Targeted Locus by PCR

14. Dilute the extracted DNA to a final concentration of 100–200 ng/μl with ddH2O.
15. Set up a 20 μl PCR reaction using the 2× FastPfu PCR SuperMix (see note 5) as follows:

2× FastPfu PCR SuperMix	10 μl
Forward primer	1 μl
Reverse primer	1 μl
DNA template	1 μl
H$_2$O	7 μl

16. Perform a PCR with the following cycling conditions (see note 6):

    1 cycle: 98 °C for 2 min (initial denaturation)
    35 cycles: 98 °C for 10 s (denaturation)
    62 °C for 20 s (annealing)
    72 °C for 1 min (extension)
    1 cycle: 72 °C for 2 min (final extension)

17. Run 2–5 μl of the PCR product on a 1 % agarose gel to verify the amplification.
18. Purify the PCR product with the QIAquick PCR purification kit according to the manufacturer's instructions (see note 7).

8.3.2.5 Digestion of the DNA Heteroduplex with T7E1

19. Mix 10 μl of the PCR product, 1 μl T7E1, 2 μl 10× T7E1 buffer, and 7 μl H_2O to a total volume of 20 μl.
20. Incubate for 30 min at 37 °C.
21. Load 10 μl of the sample with loading buffer on a 2 % agarose gel.
22. Run the gel until the loading buffer (blue dye) has migrated to the bottom of the gel.
23. Capture the gel image with a UV imaging station.

8.3.2.6 Calculation of the Cleavage Efficiency

24. Save the gel image in the TIFF format that can be opened in ImageJ.
25. Measure the integrated intensity of the PCR amplicon and cleaved bands by using ImageJ.
26. Estimate cutting efficiency using the following equation:

$$\text{indel}\% = 100 \times (1 - (1 - f_{cut})1/2), \quad f_{cut} = (b + c)/(a + b + c)$$

where a is the integrated intensity of the undigested PCR product and b and c are the integrated intensities of each cleavage product.

8.3.3 Design and Clone Donor Plasmid Based on the Best sgRNA

1. Retrieve the sequence of 1 kb surrounding OCT4 stop codon by using Genome Browser (http://genome.ucsc.edu/cgi-bin/hgBlat?command=start).
2. Conduct primary PCR to obtain the homology arm sequence.

Design primers by Primer3Plus (http://primer3plus.com/cgi-bin/dev/primer3-plus.cgi). Set the product size ranges as 1.0–1.5 kp. Make sure that there is at least 400 bp at both left and right homology arms.

3. Double-check PCR product by using in silico PCR tool at Human Genome Browser Gateway. (http://genome.ucsc.edu/cgi-bin/hgPcr?hgsid=425024239_nHNy7CS8MQDKSKIRALRKaq4r39ns). Make sure there is only product at OCT4 locus.

4. Conduct PCR using KAPA HiFi polymerase, using the following cycling conditions: 98 °C for 2 min, 1 cycle; 98°C for 20 s, 64 °C for 15 s, 72 °C for 30 s, 30 cycles. Purify the PCR products with the QIAquick PCR purification kit and confirm the product by Sanger sequencing.

5. Conduct secondary PCR to generate 3 elements of knock-in donor template with ∼ 25-bp overlapping sequence: (1) left HR arm, (2) E2A-GFP, and (3) right HR arm. In the primer design, add the sgOCT4 recognition sequence, including the NGG PAM at the 5′ end of the left HR arm and at the 3′ end of the right HR arm. If the sgRNA targets the OCT4 ORF sequence, introduce 2 sense mutations at the sgRNA target site (see note 8).

6. Conduct PCR using KAPA HiFi polymerase, using the following cycling conditions: 98 °C for 2 min, 1 cycle; 98 °C for 20 s, 64 °C for 15 s, 72 °C for 15 s, 4 cycles; 98 °C for 20 s, 68 °C for 15 s, 72 °C for 15 s, 20 cycles. Purify the PCR products with the QIAquick PCR purification kit.

7. Conduct tertiary PCR to assemble the 3 pieces together. Use the KAPA HiFi polymerase and the following cycling conditions: 98 °C for 2 min, 1 cycle; 98 ° C for 20 s, 64 °C for 15 s, 72 °C for 30 s, 4 cycles; 98 °C for 20 s, 68 °C for 15 s, 72 °C for 30 s, 20 cycles. Purify the PCR products with the QIAquick PCR purification kit.

8. Clone the PCR product into pJET1.2/blunt cloning vector using CloneJET PCR Cloning Kit. Set up the following ligation reaction on ice:

2× reaction buffer	10 μl
Non-purified PCR product from step 4	1 μl
pJET1.2/blunt cloning vector	1 μl
Water, nuclease-free up to	19 μl
T4 DNA ligase	1 μl

Vortex briefly and centrifuge for 3–5 s. Incubate the ligation mixture at room temperature for 5–30 min. Use the ligation mixture directly for transformation. Incubate the E. coli overnight at 37 °C. Pick 3 clones into 1–2 ml LB medium and shake it vigorously at 37 °C for overnight. The next day, isolate the plasmid DNA by using a QIAprep Spin Miniprep Kit according to the manufacturer's instructions. Verify the sequence of each clone by using Sanger sequencing. Isolate the correct clone using Qiagen Plasmid Plus Maxi Kit (see note 9).

8.3.4 Creating the OCT4-GFP Reporter iPSC Line

8.3.4.1 Preparation of iPSCs

1. Prepare a 50 ml aliquot of E8 medium supplemented with 10-μM ROCK inhibitor.
2. Coat the 6-well tissue culture plate with 1 ml diluted Matrigel 1:50 in cold DMEM/F12 medium and place the plate at 4 °C overnight or at least 30 min at 37 °C before use.
3. Thaw a vial of cells by gently agitating them around in a 37 °C water bath, transfer the cells to a 15-ml tube, add 5 ml of E8 medium, and centrifuge at 400 g for 5 min at room temperature.
4. Aspirate the supernatant, and seed the cells in a Matrigel-coated 6-well plate. Culture the cells with 2 ml prewarmed E8 medium containing 10-μM ROCK inhibitor.
5. Replace with E8 medium without ROCK inhibitor the next day and refeed cells with fresh E8 medium daily.
6. Passage the cells two days before transfection using Accutase at an appropriate density to achieve roughly 70–80 % confluency in 48 h.

8.3.4.2 Nucleofection of iPSCs

7. iPSCs were cultured in medium supplemented with 10-μM ROCK inhibitor 2–4 h before electroporation.
8. Wash the cells in the 6-well plate with 1–2 ml prewarmed PBS.
9. Add 0.5 ml/well prewarmed Accutase to the digest the iPSCs for 3–5 min at 37 °C.
10. Add 2× volume of prewarmed PBS and pipet gently to dissociate the iPSCs into single cells.
11. Count the number of cells using hemacytometer.
12. Centrifuge the cells at 200 g for 5 min at room temperature.
13. Remove the supernatant and resuspend 8×10^5 iPSCs in 100 μl Human ES Cell Nucleofector Solution with 4 μg of sgOCT4-Cas9 plasmid and 4 μg of GFP donor plasmid.
14. Transfer cells to an electroporation cuvette. Nucleofect the cells with the B-016 program.
15. Add prewarmed 0.5 ml E8 medium supplemented with 10-μM ROCK inhibitor into the cuvette immediately after nucleofection, gently mix the nucleofected cells using the plastic pipet provided with the Nucleofector Kit (see note 10).
16. Transfer the nucleofected cells to one well of the Matrigel-coated 6-well plate. Culture the cells with 2 ml E8 medium containing 10-μM ROCK inhibitor.

17. Grow the cells overnight at 37 °C.
18. Refeed the cells with fresh E8 medium without ROCK inhibitor on the next day after nucleofection.

8.3.4.3   Single-Cell Sorting of GFP-Positive iPSCs by FACS

19. Three to four days after nucleofection, when the cells have reached 70–80 % confluency, dissociate the cells with 0.5 ml of 0.5 mM EDTA buffer at 37 °C for 5 min.
20. Remove the EDTA bufferand dissociate the cells by adding 600 μl of E8 medium, and dissociate the cells by gentle pipetting.
21. Transfer the resuspended cells into a 5-ml tube and centrifuge the cells at 200 g for 5 min at room temperature.
22. Aspirate out the medium and resuspend the cells in 200–400 μl of FACS medium.
23. Filter the cells into a 5-ml FACS tube and put the cells on ice until sorting.
24. Coat the 96-well plate with 50 μl Matrigel per well at 4 °C overnight and add 100 μl E8 medium containing 10-μM ROCK inhibitor before sorting (see note 11).
25. GFP-positive iPSCs were sorted into 96-well plates at 1 cell per well on the BD FACSAria.
26. Grow the cells at 37 °C by changing the E8 medium every 2–3 days until reaching 10–20 % confluency.
27. Expand the cells in the 96-well plates into 2 sets of 24-well plates until reaching 80 % confluency.

8.3.4.4   Confirmation of Gene Knock-in by Junction PCR

28. Harvest cells in one set of 24-well plates for DNA extraction.
29. Dissociate the cells using 0.5 mM EDTA buffer and centrifuge the cells at 200 g for 5 min at room temperature.
30. Remove the medium and use the Genomic DNA Extraction Kit (Qiagen) to extract genomic DNA.
31. Dilute the genomic DNA to a final concentration of 100–200 ng/μl with H_2O.
32. Design a pair of PCR primers to amplify the junction between OCT4 genome and inserted GFP donor. Calculate the length of the PCR product.
33. Set up a 20-μl PCR reaction using the 2× FastPfu PCR SuperMix as follows:

2× FastPfu PCR SuperMix	10 μl
Forward primer	1 μl
Reverse primer	1 μl
DNA template	1 μl
H_2O	7 μl

34. Perform a PCR with the following cycling conditions:

    1 cycle: 98 °C for 2 min (initial denaturation)
    35 cycles: 98 °C for 10 s (denaturation)
    62 °C for 20 s (annealing)
    72 °C for 1 min (extension)
    1 cycle: 72 °C for 2 min (final extension)

35. Run the PCR products on a 1 % agarose gel to verify the amplification.
36. Pick the clones with the corrected knock-in for further culture.

8.4   Notes

1. The specificity of the Cas9 nuclease is determined by the 20-nt guide sequence within the sgRNA. Our unpublished data and other investigators show that 17- or 18-nt truncated sgRNA improves the target specificity without affecting the targeting efficiency.
2. A previous study shows that the nucleotide just in front of NGG significantly affects the targeting efficiency of the sgRNA. If the nucleotide is G, the targeting efficiency is highest, followed by A, C, and T.
3. Occasionally, BbsI does not cut the plasmid efficiently. In this case, before the enzyme digestion, treat the DNA at 60 °C for 5 min followed by cooling down on ice for 1 min.
4. To achieve high transfection efficiency, the cells should be evenly distributed. Two approaches may help prevent the clustering of cells in the center of the 24 wells: (1) Rock the plate front to back twice, right before placing it back into the incubator; (2) decrease the volume of culture medium from 500 to ∼ 300 μl.
5. To minimize error in the amplified PCR product, it is important to use a high-fidelity polymerase. Other high-fidelity enzymes, such as KAPA HiFi (Kapa Biosystems) or Phusion High-Fidelity DNA Polymerases (NEB) may be used.
6. Generally, a step of 4 °C for ∞ is unnecessary for the PCR reaction.
7. If there are no primer dimers or extra bands, it is unnecessary to purify the PCR product.
8. Introducing the sgRNA targeting sequences flanking the knock-in donor DNA increases the knock-in efficiency by fivefold or more.

9. For pJET cloning, the volume of reaction can be reduced to a total volume of 3 μl.
10. Treating iPSCs with ROCK inhibitor before and after nucleofection is critical to minimize cell death.
11. Sorting iPSCs into well plates pre-seeded with MEF feeder cells may further increase survival of iPSCs.

References

1. Zhou T, Benda C, Dunzinger S, et al. Generation of human induced pluripotent stem cells from urine samples. Nat Protoc. 2012;7(12):2080–9.
2. Chou BK, Mali P, Huang X, et al. Efficient human iPS cell derivation by a non-integrating plasmid from blood cells with unique epigenetic and gene expression signatures. Cell Res. 2011;21(3):518–29.
3. Su RJ, Baylink DJ, Neises A, et al. Efficient generation of integration-free ips cells from human adult peripheral blood using BCL-XL together with Yamanaka factors. PLoS ONE. 2013;8(5):e64496.
4. Liu SP, Li YX, Xu J, et al. An improved method for generating integration-free human induced pluripotent stem cells. Zhongguo shi yan xue ye xue za zhi / Zhongguo bing li sheng li xue hui = J exp hematol Chin Assoc Pathophysiol. 2014;22(3):580–7.
5. Su RJ, Neises A, Zhang XB. Generation of iPS cells from human peripheral blood mononuclear cells using episomal vectors. Methods Mol Biol. 2014.
6. Chou BK, Gu H, Gao Y, et al. A facile method to establish human induced pluripotent stem cells from adult blood cells under feeder-free and xeno-free culture conditions: a clinically compliant approach. Stem cells trans med. 2015;4(4):320–32.
7. Kim K, Doi A, Wen B, et al. Epigenetic memory in induced pluripotent stem cells. Nature. 2010;467(7313):285–90.
8. Hsu PD, Lander ES, Zhang F. Development and applications of CRISPR-Cas9 for genome engineering. Cell. 2014;157(6):1262–78.
9. Cox DB, Platt RJ, Zhang F. Therapeutic genome editing: prospects and challenges. Nat Med. 2015;21(2):121–31.
10. Mali P, Yang L, Esvelt KM, et al. RNA-guided human genome engineering via Cas9. Science. 2013;339(6121):823–6.
11. Cong L, Ran FA, Cox D, et al. Multiplex genome engineering using CRISPR/Cas systems. Science. 2013;339(6121):819–23.
12. Fu Y, Sander JD, Reyon D, Cascio VM, Joung JK. Improving CRISPR-Cas nuclease specificity using truncated guide RNAs. Nat Biotechnol. 2014;32(3):279–84.
13. Yang H, Wang H, Shivalila CS, Cheng AW, Shi L, Jaenisch R. One-step generation of mice carrying reporter and conditional alleles by CRISPR/Cas-mediated genome engineering. Cell. 2013;154(6):1370–9.
14. Wang H, Yang H, Shivalila CS, et al. One-step generation of mice carrying mutations in multiple genes by CRISPR/Cas-mediated genome engineering. Cell. 2013;153(4):910–8.
15. Li H, Haurigot V, Doyon Y, et al. In vivo genome editing restores haemostasis in a mouse model of haemophilia. Nature. 2011;475(7355):217–21.
16. Yin H, Xue W, Chen S, et al. Genome editing with Cas9 in adult mice corrects a disease mutation and phenotype. Nat Biotechnol. 2014;32(6):551–3.
17. Wu Y, Liang D, Wang Y, et al. Correction of a genetic disease in mouse via use of CRISPR-Cas9. Cell Stem Cell. 2013;13(6):659–62.

18. Schwank G, Koo BK, Sasselli V, et al. Functional repair of CFTR by CRISPR/Cas9 in intestinal stem cell organoids of cystic fibrosis patients. Cell Stem Cell. 2013;13(6):653–8.
19. Li K, Wang G, Andersen T, Zhou P, Pu WT. Optimization of genome engineering approaches with the CRISPR/Cas9 system. PLoS ONE. 2014;9(8):e105779.
20. Cristea S, Freyvert Y, Santiago Y, et al. In vivo cleavage of transgene donors promotes nuclease-mediated targeted integration. Biotechnol Bioeng. 2013;110(3):871–80.
21. Byrne SM, Ortiz L, Mali P, Aach J, Church GM. Multi-kilobase homozygous targeted gene replacement in human induced pluripotent stem cells. Nucleic Acids Res. 2015;43(3):e21.
22. Chu VT, Weber T, Wefers B, et al. Increasing the efficiency of homology-directed repair for CRISPR-Cas9-induced precise gene editing in mammalian cells. Nat biotechnol. 2015.
23. Maruyama T, Dougan SK, Truttmann MC, Bilate AM, Ingram JR, Ploegh HL. Increasing the efficiency of precise genome editing with CRISPR-Cas9 by inhibition of nonhomologous end joining. Nat Biotechnol. 2015.
24. Fu Y, Foden JA, Khayter C, et al. High-frequency off-target mutagenesis induced by CRISPR-Cas nucleases in human cells. Nat Biotechnol. 2013;31(9):822–6.
25. Veres A, Gosis BS, Ding Q, et al. Low incidence of off-target mutations in individual CRISPR-Cas9 and TALEN targeted human stem cell clones detected by whole-genome sequencing. Cell Stem Cell. 2014;15(1):27–30.
26. Smith C, Gore A, Yan W, et al. Whole-genome sequencing analysis reveals high specificity of CRISPR/Cas9 and TALEN-based genome editing in human iPSCs. Cell Stem Cell. 2014;15 (1):12–3.

Chapter 9
Humanized Mouse Models with Functional Human Lymphoid and Hematopoietic Systems Through Human Hematopoietic Stem Cell and Human Fetal Thymic Tissue Transplantation

Zheng Hu, Feng Jin, Bing Chen, Jinglong Guo, Jin He, Zhigang Liu, Bin Liu and Yong-Guang Yang

Abstract In order to better translate laboratory findings into clinical practice, numerous efforts have been made to develop humanized mouse models with functional human lymphoid and hematopoietic systems, which possess far more advantages than conventional animal models. Among several humanized mouse models, one established by transplantation with human hematopoietic stem cells together with human fetal thymic tissues has reconstituted nearly all of the lineages of the human functional lymphoid and hematopoietic cells. This humanized mouse model has been broadly used to study human physiology and pathology and to evaluate new therapeutic drugs for diseases relevant to the human lymphoid and hematopoietic system such as human immunodeficiency virus (HIV), hepatitis B virus (HBV), cancer, plasmodium malaria, and others. In this chapter, we introduce the method to establish this humanized mouse model in detail based on our experiences.

Keywords Humanized mouse · Hematopoiesis · Hematopoietic stem cell · Immune reconstitution · Thymus

Z. Hu (✉) · F. Jin · B. Chen · J. Guo · J. He · Z. Liu · B. Liu · Y.-G. Yang (✉)
First Hospital of Jilin University, Changchun, China
e-mail: zhenghu0108@163.com

Y.-G. Yang
e-mail: yy2324@columbia.edu

Y.-G. Yang
Columbia Center for Translational Immunology, Columbia University College of Physicians and Surgeons, New York, NY, USA

© The Author(s) 2015
T. Cheng (ed.), *Hematopoietic Differentiation of Human Pluripotent Stem Cells*,
SpringerBriefs in Stem Cells, DOI 10.1007/978-94-017-7312-6_9

9.1 Introduction

The requirement for humanized mouse models is largely driven by the fact that many findings from conventional animal models cannot be applied to humans due to the evolutionary differences among species. Without the help of humanized mice, some pathogens that infect human cells specifically, such as human immunodeficiency virus (HIV) and hepatitis B virus (HBV), cannot be studied directly in vivo. Humanized mice with functional human lymphoid-hematopoietic systems are particularly valuable in the study of these pathogens and therapies that specifically target human immune cells. Over the last two decades, a number of humanized mouse models have been created for the study of human immune responses in vivo [1]. Among these models, two are widely used and of particular mention. One is created by the injection of human CD34$^+$ hematopoietic stem/progenitor cells (HSCs/HPCs) into neonatal immunodeficient mice [2, 3]. Although transplantation of human HSCs/HPCs has been inefficient in achieving human T-cell development in adult immunodeficient mice [4, 5], the humanized mice with human T cells differentiated from the recipient mouse thymus can be achieved by transplantation of human CD34$^+$ cells into newborn immunodeficient mice [2, 3]. However, the thymus in these mice remained small with a cellularity of approximately 3×10^5, which is less than 1 % of an immunocompetent mouse thymus, indicating inefficient human thymopoiesis in the mouse thymus [3]. Furthermore, although such humanized mice can generate in vivo human immune responses to viral infections (e.g., EBV) [3], there is increasing evidence of the failure of these mice to develop efficient antigen-specific immune responses [6–9]. In addition, these mice cannot produce T cell-dependent antibodies after immunization with antigens [9, 10]. This is probably due to the lack of human leukocyte antigen (HLA)-restricted antigen recognition by human T cells as improved antigen-specific human T cell and antibody responses were observed in HLA-transgenic mice receiving human CD34$^+$ cells [6–8, 10]. The thymus is also important for the development of the major histocompatibility complex (MHC) restriction of human regulatory T cells (Tregs) [11]. Moreover, thymic stromal lymphopoietin (TSLP) expressed by thymic epithelial cells within the Hassall corpuscles is critical for the generation of human natural CD4$^+$CD25$^+$Foxp3$^+$ Tregs [12]. Thus, although CD4$^+$Foxp3$^+$ cells are detectable in newborn mice receiving CD34$^+$ cells [13], further studies are needed to assess the function of human Tregs in these mice as hematopoietic cells do not express TSLP [14] and mouse TSLP does not have cross-reactivity with human cells [15]. The other humanized mouse model is created by co-transplantation of human fetal thymic tissues and CD34$^+$ cells [16], which is discussed and illustrated in this chapter.

Human HSC xenotransplantation is required for creating humanized mice with human hematopoietic and lymphoid systems (referred to as HHLS mice [1, 16]). NOD/SCID mice and their derivatives are the preferred xenogeneic recipients for humanized mouse models because, in addition to deficiency in T and B cells, they have partially impaired NK cell and macrophage function [1, 17]. Furthermore, human CD47 is capable of interacting with macrophage inhibitory receptor, SIRPα, in NOD mice, which inhibits macrophage-mediated rejection of human hemato-poietic cells [18]. Co-transplantation of human CD34$^+$ fetal liver cells (i.v.) and human fetal thymic/liver grafts (under the renal capsule) in NOD/SCID mice or their derivatives can establish humanized mice (we named Thy/HSC mice) with high levels of human lymphohematopoietic cells as shown in Fig. 9.1 [16]. Almost all major human T cell subsets, including CD4, CD8, Treg, and invariant natural killer T (iNKT) cells were detected in these mice [16, 17, 19, 20]. Human CD4 and CD8 T cells expressing the HIV co-receptors, CXC chemokine receptor type 4 (CXCR4) and CC chemokine receptor type 5 (CCR5), were also detected in the spleen and intestine of these humanized mice, confirming the suitability of these mice for the study of HIV pathogenesis [17, 21]. Furthermore, these humanized mice have structured secondary lymphoid organs repopulated by human immune cells, providing the sites for immune response (Fig. 9.1) [16, 19]. The model has been replicated and termed the "bone marrow–liver–thymus (BLT)" mouse by another group [22]. We have recently reported that implantation of human liver tissue is dispensable for the establishment of the humanized mouse model [17].

The humanized mice established by human HSC/Thymus transplantation can generate functional in vivo immune responses. High levels of human IgM and IgG antibodies can be detected in the sera of Thy/HSC mice, indicating that the mice have functional human B cells [16]. Moreover, in vivo immunization with 2,4-dinitrophenyl hapten-keyhole limpet hemocyanin (DNP23-KLH) can induce strong antigen-specific T cell responses and the T cell-dependent secretion of DNP-specific human IgG (including IgG1, IgG2, and IgG3) antibodies (Fig. 9.2),

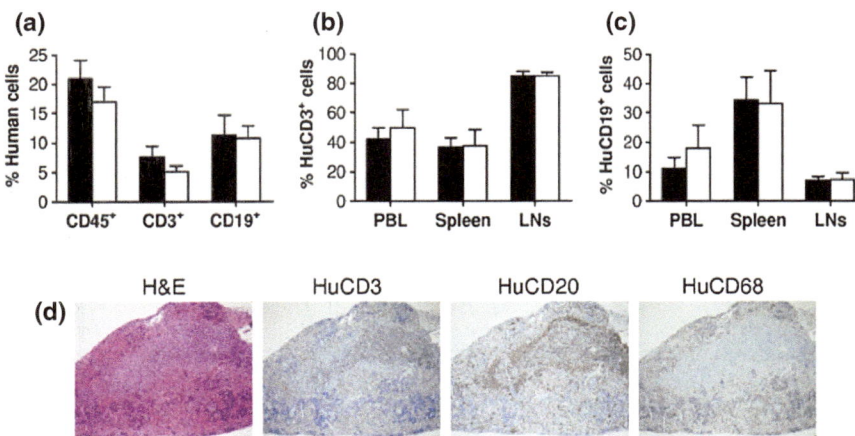

Fig. 9.1 Human immune cell reconstitution in human mice co-transplantation of human fetal thymic tissue and Human CD34$^+$ FLCs. **a** Levels of total human lymphohematopoietic CD45$^+$ cells, CD3$^+$ T cells, and CD19$^+$ **b** cells in PBMCs were analyzed by flow cytometry at week 9 after human tissue/cell transplantation. (*filled square*) and (*open square*) represent hu-mice that with ($n = 7$) or w/o ($n = 5$) DNP23-KLH immunization. **b, c** Levels of human CD3$^+$ cells (**b**) and human CD19$^+$ cells (**c**) in PBMCs, spleen, and LNs of hu-mice analyzed by flow cytometry at time of death (at around 20 weeks after humanization). Error bars represent SEM. **d** White pulp formation in hu-mouse spleen. Shown are sections prepared from a representative hu-mouse spleen stained with hematoxylin and eosin, anti-human CD3, CD20, and CD68

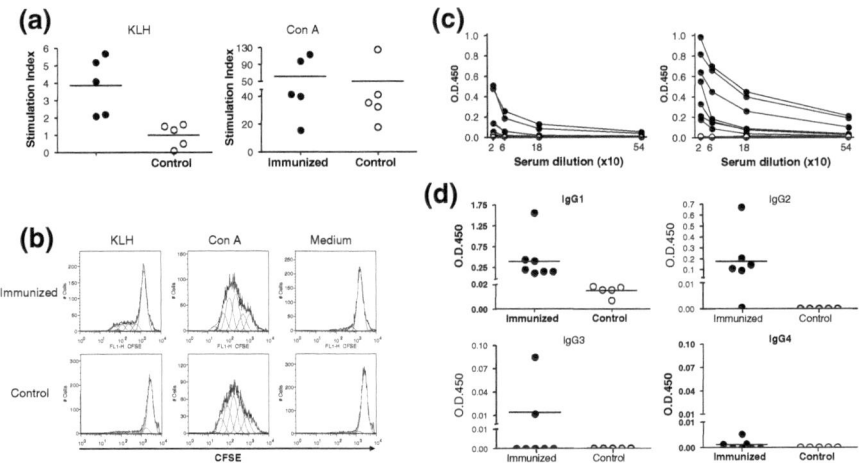

Fig. 9.2 Antigen-specific T cell and antibody responses in immunized hu-mice. **a** Proliferation of human CD3⁺ T cells in response to KLH (*left*) and Con **a** (*right*). Stimulation index of each individual hu-mouse in DNP23-KLH-immunized (*filled circle*) and control (*open circle*) groups are shown. **b** Representative histograms showing CFSE levels in gated human CD3⁺ T cells from DNP23-KLH-immunized (*top*) and control (*bottom*) groups. Cells stimulated with KLH, Con **a**, and medium are shown in *left*, *middle*, and *right* panels, respectively. **c** Serum levels of DNP-specific IgG in DNP23-KLH-immunized (*filled circle*) and PBS control (*open circle*) mice at week 1 after booster immunization (*left*) and at time of death (i.e., 2 or 4 weeks after booster immunization; *right*). Each symbol represents an individual hu-mouse. **d** Serum levels of DNP-specific human IgG1, IgG2, IgG3, and IgG4 in DNP23-KLH-immunized (*filled circle*) and PBS control (*open circle*) hu-mice at time of death (i.e., 2 or 4 weeks after booster immunization). Each symbol represents an individual mouse. Horizontal lines in **a**, **d** represent mean values

demonstrating that functional DC–T cell and T–B cell interactions exist in these humanized mice [19]. Robust HLA class I- and class II-restricted adaptive immune responses were also detected after infection with the Epstein–Barr virus (EBV) [22]. Similarly, humanized Thy/HSC mice were found to generate HIV-specific CD4 and CD8 T cells and antibody responses following HIV infection; thus, these mice are considered to be one of the best models to study HIV infection and evaluate new drugs for HIV therapy in vivo [1, 21]. Thy/HSC mice are able to mediate robust rejection of allogeneic (such as allogeneic ESC derived teratomas) and xenogeneic (such as porcine skin, islet and thymus) grafts [16, 23, 24], providing a valuable animal model to investigate the regimen to induce tolerance in the settings of allogeneic and xenogeneic transplantations.

Humanized mice generated by transplantation of human hematopoietic stem cells and fetal thymic grafts cannot only be used to study human lymphoid system but also the human hematopoietic system, such as red blood cells (RBCs) and platelets [1]. Although significant human erythropoiesis and thrombopoiesis occur in the bone marrow, there are almost no mature human RBCs or platelets in the peripheral blood of humanized mice. Our recent studies showed that rapid rejection mediated by

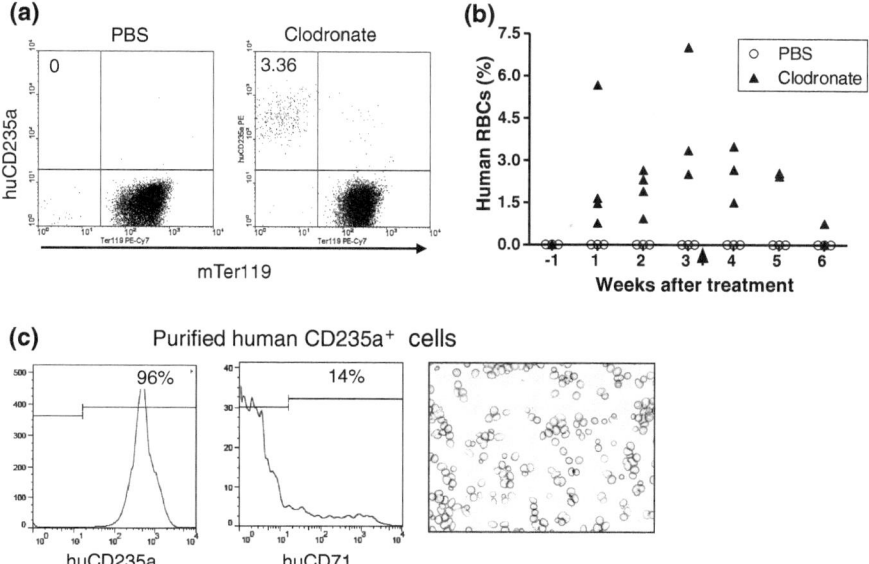

Fig. 9.3 Reconstitute of mature human red blood cells in humanized mice after macrophage depletion. Humanized NOD/SCID mice were treated at 13 weeks after human CD34+ fetal liver cell transplantation with clodronate liposomes ($n = 6$) or PBS liposomes ($n = 3$, 100 μl at days 0, 2, 7, 12, 17, and 22). Blood was collected 1 week before and weekly after clodronate liposome treatment, and the percentages of human RBCs were determined by flow cytometry. **a** Representative flow cytometric profiles at week 3 after treatment. **b** Percentages of human CD235a$^+$ RBCs in blood at the indicated time points after treatment. *Arrow* indicates the time we stopped treatment. **c** Human CD235$^+$ cells were purified by cell sorting from blood of macrophage-depleted humanized mice and analyzed by flow cytometry and cytospin. Shown are flow cytometric profiles of human CD235a and CD71 expression and Wright Giemsa staining of the purified cells

NOD/SCID macrophages is a critical factor leading to the deficiency of human RBCs and platelets in humanized mouse blood [25, 26]. As shown in Fig. 9.3, CD235a$^+$CD71$^-$ enucleated mature human RBC can be generated in humanized mice with human lymphohematopoietic cell reconstitution after macrophage depletion by clodronate liposomes [25]. Furthermore, mouse macrophage depletion also results in the full restoration of human platelets in humanized mice (Fig. 9.4) [26].

9.2 Materials

9.2.1 Mouse

NOD/SCID mice, NOD·Cg-Prkdc^{scid} Il2rg^{tm1Wjl}/SzJ (NOD/SCID/γc−/− or NSG) mice are purchased from the Jackson Laboratory (Bar Harbor, ME) and housed in a specific pathogen-free micro-isolator environment. Mice are used in experiments at 6–8 weeks of age.

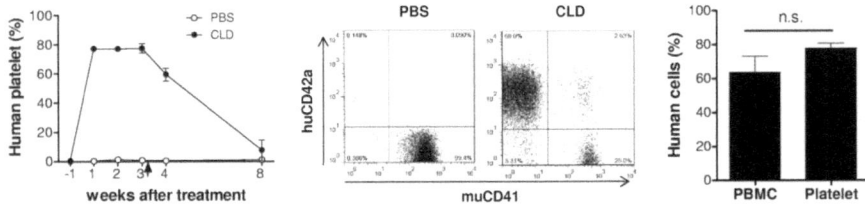

Fig. 9.4 Full restoration of human platelets in humanized mice after mouse macrophage depletion. Thirteen-week-humanized NOD/SCID mice were treated with clodronate PBS (CLD) (*n* = 4) or PBS liposomes (*n* = 3; 100 μl at days 0, 2, 7, 12, 17, and 22). Blood was collected 1 week before and at several time points after treatment, and the levels of human platelets and human PBMCs were determined by flow cytometry. Shown are levels (mean ± SEMs) of human platelet chimerism at the indicated times (*left*), representative flow cytometry profiles at week 3 after treatment (*middle*), and percentages (mean ± SEM) of human platelet versus CD45+ PBMC chimerism at week 3 after treatment (*right*). NS indicates not significant

9.2.2   Fetus

Human fetal thymus and liver tissues at gestational ages of 17–20 weeks are used to make humanized mice. Protocols involving the use of human tissues and animals should be approved by the Human Research Committee and Subcommittee on Research Animal Care of the institute.

9.2.3   Human Hematopoietic Stem Cell Purification

CD34 MicroBead Kit (Miltenyi Biotec, Auburn, CA); LS column: (Miltenyi Biotec, Auburn, CA); Bone marrow medium: 500 ml Media 199 (Gibco) + 5 ml Hepes Buffer + 5 ml DNAse (1 mg/ml) + 40μl gentamycin (50 mg/ml); MACS buffer: 5 % human AB serum, 2 mM EDTA in PBS; Histopaque®-1077 (Sigma); 70-μm cell strainer (BD Falcon); 40-μm cell strainer (BD Falcon).

9.2.4   Materials for Human Fetal Thymic Graft Transplantation

9.2.4.1   Anesthetic Agent

0.8 ml of ketamine HCl (100 mg/ml NDC 0409-2051-05) and 0.6 ml Xyla-Ject (20 mg/ml NDC 57319-362-26) in 10 ml of sterilized PBS.

9.2.4.2   Surgical Instruments for Human Fetal Thymic Graft Transplantation

Micro-dissecting scissors	Roboz	RS-5673
Genuine Dumont tweezers	Roboz	RS-4968
Genuine Dumont tweezers	Roboz	RS-4972
Genuine Dumont tweezers	Roboz	RS4978

9.2.5   Flow Cytometric Analysis

9.2.5.1   Fluorescent-Conjugated mAbs and Isotype-Matched Control Abs

Anti-human CD45, CD19, CD3, CD4, CD8, CD11c, CD14, CD123, CD235a, CD42a, HLA-DR; anti-mouse CD45, CD41, and Ter119; and isotype control mAbs (All antibodies purchased from BD PharMingen, San Diego, CA). Analysis was performed on a FACS Canto or LSR II (Becton Dickinson, Mountain View, CA). Dead cells were excluded from the analysis by gating out low forward scatter and high-propidium iodide-retaining cells.

9.2.5.2   FACS Medium

100 ml 10× Hanks balanced salt solution + 900 ml D.I. H_2O + 210 μl of 1 M sodium hydroxide (NaOH) + 1 mg sodium azide (NaN_3) + 1 mg bovine serum albumin (BSA).

9.2.6   Macrophage Depletion Regimen

Clodronate liposomes can be requested from Dr. Nico Van Rooijen (http://www.clodronateliposomes.org) or prepared as described [27].

9.3   Methods

9.3.1   Human CD34 + Fetal Liver Cell (FLC) Purification and Cryopreservation

1. Smash human fetal liver pieces by 70-μm cell strainers and suspend them into 50–100 ml bone marrow medium.
2. Incubate human fetal liver cells with bone marrow medium on ice for 1 h.
3. Wash the human fetal liver cells once and suspend in 50 ml PBS.
4. Layer the human fetal liver cell suspension onto Histopaque®-1077.
5. Centrifuge the cells at 300 g for 30 min at room temperature.
6. Collect the cells found in the intermediate layer.
7. Wash the intermediate layer cells once with MACS buffer.
8. Re-suspend the cells in 10 ml of MACS buffer.
9. After dilution with trypan blue, count the live cells under the microscope.
10. Spin down the human fetal liver cells and suspend in MACS buffer (10^8 cells for 500 ul).
11. Add anti-CD34 micro-beads (100 μl/10^8 FLCs) and incubate on ice for 30 min.
12. Wash the cells once with 30 ml of MACS buffer.
13. Suspend the cells in 5 ml MACS buffer and filter with 40-μm strainer to remove cell clots.
14. Place an LS column in a suitable MACS separator and rinse once with 3 ml MACS buffer.
15. Apply the anti-CD34 micro-bead-labeled human fetal liver cells to the rinsed LS column.
16. Wash the LS column 3 times with 3 ml of MACS buffer.
17. Remove LS column from the separator and place on a suitable collection tube. Add 5 ml of MACS buffer to immediately flush out the magnetically labeled cells.
18. Split 2×10^5 cells and analyze the enrichment by performing FACS after staining cells with anti-human CD34 fluorescent antibodies (use a different clone than the anti-human CD34 micro-bead antibody).
19. Suspend the rest of the human CD34$^+$ FLCs in FBS (2×10^6 cells/ml) and mix gently with the same volume of cryopreservation medium (20 % DMSO + 80 %FBS). Aliquot the cells into cryopreservation tubes.
20. Place the tubes into cell-freezing boxes containing isopropanol and place in a −80 °C freezer overnight.
21. Preserve cells into liquid nitrogen tank for late use.

9.3.2 Human Fetal Thymic Tissues Preparation

1. Place all human fetal thymus in a 10-cm plastic dish with complete 1640 medium and keep on ice.
2. Cut the human fetal thymus tissue into 1-mm^3 pieces with micro-dissecting scissors (Roboz).
3. Put the human fetal thymic tissue into cryopreservation medium (10 % DMSO + 90 % FBS) at 5 pieces/ml.
4. Preserve the human fetal thymic tissues in liquid nitrogen for later use.

9.3.3 Humanization of NOD/SCID Mice and Their Derivatives

1. Treat NOD/SCID mice with 2–3 Gy total body irradiations (X-ray or γ-ray).
2. Administer anesthetic to NOD/SCID mice more than four hours after irradiation with 10 μl/g of anesthetic agent by i.p. injection.
3. Shave the abdomen hair of the animals.
4. Restrain animal on operating table.
5. Sterilize the animal with iodine and wipe with 70 % alcohol pads.
6. Under a magnifying glass, open the abdominal skin with scissors.
7. Open the abdominal membrane with scissors and keep open with a retractor.
8. Exteriorize the kidney with wet cotton (wet by PBS).
9. Aspirate one piece (around 1 mm^3) of thymus and liver into a hooked needle connected to a 1-ml syringe and put aside.
10. Pick up the kidney membrane with forceps and make a small hole by forceps with right hand.
11. Keep the hole opened with the forceps and make a small hole to a big pocket by the rounded tip.
12. Insert the hooked needle into the pocket and flush the thymus/liver.
13. Stitch the abdominal membrane closed with suture silk.
14. After the animal wakes up, administer an intravenous injection of 1–2 × 10^5 CD34$^+$ cells/mouse.

9.3.4 Follow the Human Lymphoid Cell Reconstitution of Humanized Mice Every 2–3 Weeks

1. Warm the animals for 5 min with a 100-W electric bulb.
2. Cut the tail vein and collect 5–7 drops of blood in a heparin-treated tube.
3. Suspend the mouse blood in 1 ml of PBS.

4. Carefully layer 1 ml of mouse blood on top of 500 μl Histopaque®-1077.
5. Spin at 300 g for 30 min at room temperature.
6. Collect the central layer cells and wash once.
7. Add the following fluorescent-conjugated antibody cocktail: anti-human CD45, anti-human CD3, anti-human CD19, anti-human CD14, anti-human CD4, anti-human CD8, Ter119, and anti-mouse CD45. Incubate for 30 min and wash once with FACS medium.
8. Examine the blood samples for human cells by flow cytometry.

9.3.5 Generation of Humanized Mice with Human RBC and Platelet Reconstitution in Periphery Blood

1. Establish humanized mice with human lymphoid cell reconstitution as shown in step "3.3" (human thymic graft is dispensable for human RBC/platelet reconstitution).
2. Follow the human chimerism in peripheral blood mononuclear cells (PBMC) every 3–4 weeks by flow cytometry (as shown in step "3.4").
3. Intravenously inject 100 μl/ea clodronate liposome into successfully humanized animals with subsequent injections of 50 μl/each clodronate liposomes every 5–7 days to prevent mouse macrophage restoration.
4. Measure the concentration of human RBCs and platelets in mouse peripheral blood by flow cytometry.

9.3.5.1   Measure Human RBC Reconstitution

1. Collect 10^6 mouse whole blood cells from the tail vein.
2. Label the blood cells with anti-Ter119 and anti-human CD235a antibodies for 30 min.
3. Wash cells with 4 ml of FACS medium by centrifuging at 400 g for 5 min.
4. Repeat "step 3."
5. Evaluate the degree of human RBC chimerism by flow cytometry.

9.3.5.2   Measure Human Platelet Reconstitution

1. Collect 2–3 drops of mouse blood from the tail vein and store in heparin-treated tubes.
2. Add 5 ml FACS medium to the blood and suspend the cells.
3. Spin the cells at 400 g for 5 min.

4. Collect the supernatant.
5. Spin the supernatant at 2500 g for 15 min.
6. Add anti-mouse CD41 and anti-human CD42a antibodies.
7. Wash the cells with 5 ml FACS medium by centrifuging at 2500 g for 15 min.
8. Repeat "step 7."
9. Evaluate the degree of human platelet chimerism by flow cytometry.

Notes

1. Keeping the whole procedure of transplantation under sterilized condition is important for the success of humanization.
2. NSG mice required fewer human $CD34^+$ FLCs than NOD/SCID for effective humanization.
3. More than 10 % of human $CD45^+$ cells (which are mainly composed of human $CD19^+$ B cells) should be detected 3 weeks after transplantation or the animal will not have good human chimerism later on.
4. Since humanized mice may develop graft-versus-host syndrome more than 20 weeks after transplantation, the window for using humanized mice for human lympho-hematopoietic system-related studies is 9–20 weeks after transplantation.
5. Clodronate liposome is toxic to humanized NOD/SCID mice, and mortality of the animals is expected in long-term experiments.

References

1. Rongvaux A et al. Human hemato-lymphoid system mice: current use and future potential for medicine. Annu Rev Immunol. 2013.
2. Ishikawa F, et al. Development of functional human blood and immune systems in NOD/SCID/IL2 receptor gamma chain(null) mice. Blood. 2005;106(5):1565–73.
3. Traggiai E, et al. Development of a human adaptive immune system in cord blood cell-transplanted mice. Science. 2004;304(5667):104–7.
4. Ito M, et al. NOD/SCID/gamma(c)(null) mouse: an excellent recipient mouse model for engraftment of human cells. Blood. 2002;100(9):3175–82.
5. Shultz LD, et al. Human lymphoid and myeloid cell development in NOD/LtSz-scid IL2R gamma null mice engrafted with mobilized human hemopoietic stem cells. J Immunol. 2005;174(10):6477–89.
6. Shultz LD, et al. Generation of functional human T-cell subsets with HLA-restricted immune responses in HLA class I expressing NOD/SCID/IL2r gamma(null) humanized mice. Proc Natl Acad Sci USA. 2010;107(29):13022–7.
7. Jaiswal S, et al. Dengue virus infection and virus-specific HLA-A2 restricted immune responses in humanized NOD-scid IL2rgammanull mice. PLoS ONE. 2009;4(10):e7251.
8. Strowig T, et al. Priming of protective T cell responses against virus-induced tumors in mice with human immune system components. J Exp Med. 2009;206(6):1423–34.
9. Watanabe Y, et al. The analysis of the functions of human B and T cells in humanized NOD/shi-scid/gammac(null) (NOG) mice (hu-HSC NOG mice). Int Immunol. 2009;21(7):843–58.

10. Danner R, et al. Expression of HLA class II molecules in humanized NOD.Rag1KO. IL2RgcKO mice is critical for development and function of human T and B cells. PLoS ONE. 2011;6(5):e19826.
11. Itoh M, et al. Thymus and autoimmunity: production of CD25+ CD4+ naturally anergic and suppressive T cells as a key function of the thymus in maintaining immunologic self-tolerance. J Immunol. 1999;162(9):5317–26.
12. Watanabe N, et al. Hassall's corpuscles instruct dendritic cells to induce CD4+ CD25+ regulatory T cells in human thymus. Nature. 2005;436(7054):1181–5.
13. Jiang Q, et al. FoxP3+ CD4+ regulatory T cells play an important role in acute HIV-1 infection in humanized Rag2-/-gammaC-/- mice in vivo. Blood. 2008;112(7):2858–68.
14. Soumelis V, et al. Human epithelial cells trigger dendritic cell mediated allergic inflammation by producing TSLP. Nat Immunol. 2002;3(7):673–80.
15. Reche PA, et al. Human thymic stromal lymphopoietin preferentially stimulates myeloid cells. J Immunol. 2001;167(1):336–43.
16. Lan P, et al. Reconstitution of a functional human immune system in immunodeficient mice through combined human fetal thymus/liver and CD34+ cell transplantation. Blood. 2006;108 (2):487–92.
17. Hu Z, Yang YG. Human lymphohematopoietic reconstitution and immune function in immunodeficient mice receiving cotransplantation of human thymic tissue and CD34(+) cells. Cell Mol Immunol. 2012;9(3):232–6.
18. Takenaka K, et al. Polymorphism in Sirpa modulates engraftment of human hematopoietic stem cells. Nat Immunol. 2007;8(12):1313–23.
19. Tonomura N, et al. Antigen-specific human T-cell responses and T cell-dependent production of human antibodies in a humanized mouse model. Blood. 2008;111(8):4293–6.
20. Onoe T, et al. Human natural regulatory T cell development, suppressive function, and postthymic maturation in a humanized mouse model. J Immunol. 2011;187(7):3895–903.
21. Brainard DM, et al. Induction of robust cellular and humoral virus-specific adaptive immune responses in human immunodeficiency virus-infected humanized BLT mice. J Virol. 2009;83 (14):7305–21.
22. Melkus MW, et al. Humanized mice mount specific adaptive and innate immune responses to EBV and TSST-1. Nat Med. 2006;12(11):1316–22.
23. Rong Z, et al. An effective approach to prevent immune rejection of human ESC-derived allografts. Cell Stem Cell. 2014;14(1):121–30.
24. Tonomura N, et al. Pig islet xenograft rejection in a mouse model with an established human immune system. Xenotransplantation. 2008;15(2):129–35.
25. Hu Z, Van Rooijen N, Yang YG. Macrophages prevent human red blood cell reconstitution in immunodeficient mice. Blood. 2011;118(22):5938–46.
26. Hu Z, Yang YG. Full reconstitution of human platelets in humanized mice after macrophage depletion. Blood. 2012;120(8):1713–6.
27. van Rooijen N, Sanders A. Liposome mediated depletion of macrophages: mechanism of action, preparation of liposomes and applications. J Immunol Methods. 1994;174(1–2):83–93.